THE BILINGUAL
EDGE

THE BILINGUAL EDGE

EDGE

Why, When, and How to Teach

Your Child a Second Language

KENDALL KING, PH.D.
ALISON MACKEY, PH.D.

Collins

An Imprint of HarperCollinsPublishers

HarperCollins books may be purchased for educational, business, or sales promotional use. For information, please write: Special Markets Department, HarperCollins Publishers, 10 East 53rd Street, New York, NY 10022.

FIRST EDITION

Designed by Mia Risberg

Library of Congress Cataloging-in-Publication Data is available upon request.

ISBN: 978-0-06-124656-2
ISBN-10: 0-06-124656-5

07 08 09 10 11   WBC/RRD   10 9 8 7 6 5 4 3 2 1

CONTENTS

Introduction   vii

SECTION ONE   *Why Are Two Languages Better Than One?*

1 How Can Your Child Benefit from the Bilingual Edge?   3
2 Myths and Misconceptions about Learning a Second
  Language: What's the Real Deal?   16

SECTION TWO   *Which Language and When?*

3 Which Language Is Right for Your Child?   37
4 When Should Your Child Start Learning Another Language?   55
5 How Do Individual Differences Like Birth Order, Gender,
  Personality, Aptitude, and Learning Style Affect Your
  Child's Language Learning?   77

SECTION THREE *How?*

6 How Can You Best Promote Language
   Learning at Home?   97
7 Language Learning through Edutainment: Can Children
   Be Educated and Entertained at the Same Time?   133
8 What Are the Characteristics of Good Second Language
   Learning Programs and Teachers and How
   Can You Find Them?   152

SECTION FOUR *What If . . . ?*

9 What if My Child Mixes and Switches Languages?   183
10 What Do I Need to Know about Language Delay, So-Called
   Expert Advice, Special Needs, and My Child's Apparent
   Lack of Progress?   207
11 What Do I Need to Know about Trilingualism
   and Dialects?   220
12 What Can I Do if My Family Disagrees, My Child Resists the
   Second Language, My Family Circumstances Change, or
   if There Are Other Problems Along the Way?   235

   Conclusion: The Bilingual Edge Is for Everyone   252
   Acknowledgments   257
   References and Resources   259

INTRODUCTION

All babies are born into this world with a gift for language learning. Because of this gift, millions of young children from all around the world routinely become bilingual, and so can yours. No matter what your own language background and situation is, your child deserves—and can benefit from—all of the advantages that come with knowing more than one language. This is a special gift that you can begin to give your child from day one.

Knowing more than one language is more important today than ever before. Whether it's English, Spanish, French, Korean, or one of the other six thousand languages of the world, parents are increasingly drawn to the lifelong academic, social, cultural, and intellectual advantages that come with learning an additional language. Millions of dollars are spent each year on classes and programs, as well as CDs, DVDs, software programs, educational bilingual toys, and the like, all of which promise to help your child become bilingual. Yet many of these promises end in disappointment—in part because these products and programs are unfortunately *not* based on what science tells us about how children actually learn languages.

HOW CAN READING THIS BOOK BENEFIT YOUR CHILDREN?

The "edge" or "advantage" that comes with bilingualism is a valuable resource for everyone. This edge is evident not only through tests of intelligence and academic ability, but also in children's enhanced creativity, self-esteem, cross-cultural understanding, and future job opportunities, among other things. You'll finish this book with the knowledge, skills, and confidence you need to provide your children with this bilingual edge—an edge we know parents want for their children now more than ever before!

WHO IS THIS BOOK FOR?

The Bilingual Edge is for *all* parents who would like their children to know more than one language. This includes parents who *only* speak English, but want to introduce a foreign language effectively and efficiently to their child. This book is for parents who think learning a second language is generally a good idea, but aren't sure when to begin or which second language to choose. It's also for parents who have begun to teach their children second languages and would like to know more about what's in store for them. And it is for parents who speak a language like Spanish, Korean, or Russian and want to make sure their children are fluent bilinguals in that language *and* English.

The Bilingual Edge shows parents from all sorts of language backgrounds—with all sorts of children—how to assess and leverage their *own* families' situations for optimal child language learning. Every family, and every community, has their own unique strengths that can help in learning a second language. In the following pages, you'll discover that you don't necessarily need to fluently speak the language you want your children to learn—nor do you need to purchase an extensive (and expensive) foreign language DVD library or enroll your children in fancy, overpriced private schools. What you do need is a bit of knowledge about how language learning happens, *and* how to put this knowledge to work in your own family.

We guide you through the many questions that parents face. We provide insights from the latest research, which will help you make the best choices for your children concerning *which* language(s) to choose, *when* to introduce them, *how* best to do so, and much more. We present the latest exciting findings about second language learning, from the fields of linguistics, education, and psychology in ways that will matter and be accessible to your family. Every step of the way, key research on language learning is put into context and made relevant for you with real-life examples, activities, quick tips, and checklists, all of which clearly illustrate how you can introduce second languages to your children effectively in easy, fun, and interactive ways. Our aim is to provide scientific, reliable, research-based information and tools that can help you make the best choices for your children concerning their second language learning.

WHO ARE WE AND WHY IS THIS BOOK NEEDED?

We are professors of linguistics at Georgetown University in Washington, D.C. Kendall's area is bilingualism and Alison is an expert in second language learning. We've been colleagues and friends for more than a decade, independently writing academic articles and books *for other researchers* about how people learn and use languages. Then one year we had children—a boy and a girl, just two months apart. As we crossed into a new world of parenting (full of playgroups, parenting Web sites, books, and magazines), we realized how badly a book like *The Bilingual Edge* was needed. In particular, we noticed three things: first, the overwhelming enthusiasm and desire among many, many parents to promote early second language learning; second, widespread misunderstandings about how second languages are learned (and how they aren't); and third, a remarkable lack of unbiased, scientifically based—but understandable—information out there for parents. Time and again, we hear from parents wondering how to incorporate more than one language into their child's life. Often from experience, parents know that suffering through two years of high school French or Spanish doesn't cut it, but they don't know what the best alternatives are.

We resolved to put our heads together to write a book that helps parents put the latest research into practice with their own children. As parents trying to bring up children who speak more than one language ourselves, we want to share with you what we've learned personally and through our many years of academic experience researching second language learning (we've written almost a hundred books and articles between us, and read many more!). Aside from lots of tools and how-to information in the chapters that follow, you'll find facts and insights that strengthen your rationale and your personal resolve to bring your child up knowing more than one language. Our goal is to help our children—and yours—become fluent bilingual speakers. Starting second language learning now will give your child the best chance of becoming fluent and developing a nativelike accent.

The Bilingual Edge provides everything you need to help your child make the most of each and every opportunity for language learning and to reap all the benefits that come with bilingualism. So, let's roll up our sleeves and get started!

Why Are Two Languages Better Than One?

How Can Your Child Benefit
from the Bilingual Edge?

You probably already have the idea that you'd like to raise your child to speak more than one language. You might have begun already. And you're not alone! Many parents feel—and as parents and scientists we wholeheartedly agree—that being bilingual provides an undeniable advantage in life. For children, advanced knowledge of two languages has been shown to result in specific brain benefits, like enhanced creativity and flexibility, increased test scores, and improved literacy skills, as well as social advantages such as greater cross-cultural understanding, adaptability, and increased competitiveness on the job market down the line. Language is interwoven with who we are and how we relate to others, and many parents realize that knowing two or more languages can enhance not only their children's self-esteem and identity, but also their pride in their own heritage.

Most parents reading this book have a sense of these important advantages (which is why you picked up this book to begin with!). In this chapter, we'll review some of the most important research findings that demonstrate exactly what these bilingual advantages are. Keeping these scientific findings in mind will help you persevere in the months and years ahead. This research will also arm you with the information you will need to help motivate others and even get any skeptics on your side (including those in your own family, day-care providers, doctors, and teachers who don't know the research).

KNOWING TWO LANGUAGES GIVES CHILDREN A COGNITIVE EDGE

Many of us intuitively grasp that knowing more than one language makes us smarter in some way. And indeed, this intuition is supported by lots of research. Part of the bilingual edge is that bilinguals tend to outperform monolinguals on many different sorts of tests.

In what areas do bilinguals have an edge? First, people with advanced knowledge of more than one language seem to be more creative. How is creativity measured, you may be wondering—it seems like a pretty abstract concept. Well, most frequently by asking questions like: "How many ways could you use an empty water bottle?" On these sorts of tests, bilinguals tend to produce *more* answers and also *more creative* answers. For instance, for the water bottle question, most of us would come up with the obvious answer ("filling it with water"), but bilinguals are more likely to come up with other answers too, like "filling it with sand and making a paperweight." Overall, bilinguals outperform monolinguals on most tests like these, most of the time. Something about knowing more than one language seems to make children both more creative and what researchers describe as more mentally flexible. This type of creativity is increasingly important in today's world—and can translate into success in school and in life!

For instance, many adult bilingual authors describe their bilingualism as a source of inspiration for their writing. Different sounds, grammars, and ways of saying things can provide fresh perspectives on everyday occurrences. Prominent writers who've used their bilingualism to creative advantage include Salman Rushdie, Sandra Cisneros, Isabelle Allende, Arundhati Roy, and Junot Diaz (among *many, many* others). As German-Japanese-English trilingual writer Yoko Tawada explains, "When you make a connection between two words that lie miles apart, a kind of electricity is produced in your head. There is a flash of lightning, and that is a 'wonder-full' feeling."

So creativity and flexibility are good, but what else does being bilingual buy you? Most of the cognitive advantages stem from bilinguals' greater metalinguistic awareness, which means awareness of language as an object or system. Bilingual children are more sensitive to the fact that language is a system that can be analyzed or played with. Metalinguistic awareness is what allows us to appreciate many types of jokes, puns, and metaphors. This sounds a bit abstract, but metalinguistic awareness is also linked to important academic skills, including learning to read. Children who are more metalinguistically aware have fewer problems in becoming literate and do better on tests of reading readiness. Bilingual children are more likely than monolingual children to recognize that it's possible, for instance, for one object to have two names. This also allows them to recognize linguistic ambiguities sooner than monolinguals. Because they know two languages, bilinguals are much more sophisticated than monolinguals in terms of understanding something very important about how language works! Metalinguistic awareness is something that teachers often try to foster, because of its connection to test scores and literacy. Bilinguals automatically have an edge in this sort of knowledge.

Bilinguals also outperform monolinguals on tests that require

> **• FAST FACT •**
>
> Both creativity and sensitivity to language are more and more important for success in today's schools. These cognitive advantages can help give your child an educational edge.

them to ignore distracting information. Bilinguals are better at focusing on the required task (for instance, judging the correctness of the grammar of a sentence or counting the number of words in a sentence) while disregarding misleading, irrelevant details. This is also an important advantage in today's educational environments where there is a lot of language to work with and no shortage of distractions.

SPOTLIGHT ON RESEARCH:

In Which Bed Does the Spoon Sleep?

In her 2001 book, Ellen Bialystok, a well-known researcher on bilingualism, helped us understand why bilingual children show advantages in metalinguistic awareness. She demonstrated that bilingual children outperform monolinguals in what she calls their "cognitive control of linguistic processes."

Bialystok asked approximately 120 children, aged five to nine, to judge sentences as grammatically acceptable or grammatically unacceptable *while ignoring the meaning of the sentence.* Children were given three sentence types:

- grammatically acceptable and sensible: *In which bed does the baby sleep?*
- grammatically unacceptable but sensible: *In which bed does baby the sleep?*
- grammatically acceptable but not sensible: *In which bed does the spoon sleep?*

Bialystok found that monolingual children were more easily misled by the *meaning* of the sentence (for example, they judged sentences like "In which bed does the spoon sleep?" as ungrammatical). Bilingual children, on the other hand, were better at ignoring distracting information (in this case, the meaning of the sentence) while also correctly judging the grammar of the sentences. In a nutshell, bilinguals seem better equipped than monolinguals to focus on the grammatical task at hand.

So, to recap, bilingual children have specific advantages over monolingual children, particularly in areas like metalinguistic awareness, creativity, and the ability to control linguistic processing. However, we do need to point out a few things. First, while these advantages are important, parents shouldn't be misled to believe that bilingualism influences *every* aspect of cognition. Second, findings about bilingual advantages generally apply to children who have *advanced proficiency in the two languages.* In other words, we are not talking about children who have grasped simple skills like how to count to five in Spanish and say hello and good-bye. Very occasional exposure to a second language (for example, half an hour of TV or a short class once a week) is probably not enough for significant language learning and the associated advantages to take hold. Third, other factors, such as children's exposure to books and other literacy materials, play an important role. Exposure to print (in any language) enhances metalinguistic awareness, so for optimum benefits, be sure that print media is part of your bilingual parenting plan. (More on this in chapters 6 and 7.) Still, even taking all of these little caveats into account, the substantial

> **· FAST FACT ·**
>
> For the many cognitive advantages of bilingualism to kick in, children need to reach relatively high levels of proficiency in *both* languages.

brain benefits that have been found for bilinguals make adding an additional language early on a very good investment for your child!

LEARNING ANOTHER LANGUAGE
ENHANCES CROSS-CULTURAL UNDERSTANDING

As many of us know from experience, it's hard to understand a culture without knowing the language of that culture. And research tells us when children learn a second language they are more likely to have positive attitudes toward speakers of that language.

Second language learning programs have positive effects on cross-cultural attitudes and behaviors. This effect is particularly strong for language programs that are dual or two-way (for example, where Spanish speakers learn English and English speakers learn Spanish together)—more about these in chapter 8. Children enrolled in these language programs have more positive attitudes toward the members of the other group. These children are more likely to make friends from other language and culture groups, both within their school language programs and beyond. Importantly, they also have fewer negative stereotypes about other groups. Perhaps best of all, these benefits can last throughout a child's lifetime. With multiculturalism increasing in today's global society, greater cross-cultural understanding and sensitivity is a critically important skill to give to our children, and this can be achieved through language!

SPOTLIGHT ON RESEARCH:

Building Friendships Across Language Lines

In a 1983 evaluation of a Spanish-English two-way immersion program, researchers Cazabon, Lambert, and Hall wanted to see if children formed social groups based on ethnicity or language. Children were asked questions about their best friends—who they ate lunch with, who they would invite home, who they would choose to play games with, and who they liked to sit next to. While younger children showed some preference for friends based on ethnicity or language, by the third grade, children were equally likely to have friends from different backgrounds. Linguistic or ethnic differences were no longer a factor! After some time in the two-way program, students had come to value friends as individuals rather than as members of any particular group. In other words, children learned the very important lesson in life that what matters most deep down is not what you look like or how you talk, but who you are.

Raising your child to be bilingual potentially impacts not only your child's individual success and happiness, but also the greater community. After all, one out of five households in the United States uses a language other than English, and that number is growing each year! Many of the qualities we value for our children, such as awareness and understanding of other people and cultures, the ability to make friendships across social lines, and the skills necessary for creative problem-solving, in fact are cultivated by bilingualism.

LANGUAGE AND CULTURE

Introducing children to a second language also introduces them to a second culture in less obvious ways. Anthropologists and other researchers have argued that culture and language are inextricably linked, with some even claiming that the language(s) we speak strongly influence the ways we think. For example, languages have different ways of categorizing and organizing information through their grammars and vocabularies. These language differences, in turn, potentially shape the way we view the world. So, for instance, in English, there is only one word for "corner," while in Spanish, there are two words, *rincón* and *esquina*. *Rincón* refers to the inside of the corner. (If you were in a room you might refer to the *rincón* by the window.) *Esquina* refers to the outside of the corner. (So, for instance, both of our children went through the standard period of regularly banging into *la esquina* of the coffee table when they were learning to walk.) Children who know more than one language intuitively pick up that there is more than one way to divide up and think about our shared physical and cultural world.

So, since our language can shape the way we think about and represent things, children who learn two languages also learn that different people have different views of phenomena in the world. Bilingual children have an advantage in that it is easier for them to understand that one perspective is not better or worse than another, only different.

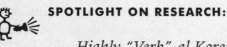

SPOTLIGHT ON RESEARCH:

Highly "Verb"-al Korean Children

Which languages we happen to learn influences not just how we see the world, but also which skills we develop. English-learning children acquire nouns (for example, *bottle*, *car*) first. However, researchers Gopnik and Choi reported in 1993 that verbs (for example, *eat, sleep*) are among the first words acquired by Korean children. This fact is probably because verbs come at the end of the sentence in Korean, and we tend to pay more attention to what comes first or last. (A quick analogy: When you've forgotten your shopping list at the grocery store, you tend to remember what was at the top and bottom, but those middle items are a mental blur.)

This language difference seems to influence children's skill development in interesting and important ways. How so? Korean children tend to perform better on tasks that are related to verbs than English speakers do. For example, Korean toddlers are better at certain verb-related tasks involving tools. But English-speaking children perform better on noun-related tasks, like categorizing objects. This difference in skill development may be linked to the different levels of attention given to nouns and verbs in the Korean and English languages. The broader message for parents is that languages put greater emphasis on different areas. Knowing more than one language potentially buys you a wider range of awareness and skills.

KNOWING MORE THAN ONE LANGUAGE ENRICHES FAMILY LIFE, CULTURE, AND COMMUNICATION

For many of the families we have worked with who have a heritage language in their background (whether they use it themselves regularly or not), being bilingual is often felt to be critical to maintaining family connections and cultural traditions. (A heritage language is one that has been spoken by previous generations and that often has some special meaning for the family.) Many parents deeply value maintaining this cultural link for their children.

And many parents also grasp the importance of the second, or heritage, language as central to maintaining cultural traditions. It's hard for children to understand or even sit still, much less participate, in an event if they don't understand the language (and, of

> **• FAST FACT •**
>
> A heritage language is spoken by previous generations of a family. It might still be used in the home or it might exist through memories of how grandparents or great-grandparents used to speak.

course, it's sometimes a challenge even if they do!). For instance, children who don't understand Korean probably aren't going to get much out of a Sunday church service in Korean. Knowledge of the heritage language is also important for participation in many local cultural celebrations and community activities. Parents know that if their children are English monolingual, they'll likely be at the margins of these events and (understandably) less interested.

Still other parents see knowledge of their heritage language as an important source of pride and self-esteem for their child. They are right! There's a lot of research evidence that children do best in school—and best in life overall—when they have a strong sense of identity and of where they come from. Many researchers have concluded that immigrant groups to the United States who maintained their cultural heritage at home—for instance, some Chinese and Indian groups—also provided their children with the strength to face challenges, and sometimes inequalities, at school. In contrast, total assimilation or loss of cultural heritage can lead to less successful outcomes at school.

Of course, for many parents, speaking the family's heritage language is not so much about the need to communicate with others immediately around them, but rather about the need to know who they are and where they came from, to provide the family with a sense of heritage and to foster closeness between parents and children. For instance, our friend Thomas is from Ghana, although he spent most of his adult life in London. His wife, Ola, is from Nigeria, but also grew up in London. The two met and married in London, but have lived in the United States for almost ten years. Their two girls, Atswei and Ayorkor, both have Ghanaian names, and

Thomas consistently uses Ga (one of Ghana's sixteen languages) with both girls. Even though there are only six hundred thousand speakers of Ga in Ghana (and far fewer in Washington, D.C.!), both Thomas and Ola feel that it is essential that the girls have some connection with Ghana as a country and with Africa more generally. If you are thinking of doing something similar with your own family, know that you are not alone! Knowledge of a heritage language can be an important part of identity.

BILINGUALISM PROVIDES
AN EDUCATIONAL AND CAREER EDGE

Leaving issues of identity and culture aside for a moment, there is no question that knowing more than one language provides an edge in both education and career achievement, more today than ever before. In today's competitive academic environment, we're all looking to give our children that additional edge. Meanwhile, global developments continue to increase the demand for bilingual professionals. Why not prepare our children for this from the get-go?

If your children are still in diapers, the job market probably seems pretty far off. Still, it's worth keeping in mind that multilingual professionals are increasingly in demand. It can be motivating both for parents and teens to know that knowledge of more than one language pays off in the job market. This trend is only likely to intensify. For example, more than 75 percent of U.S. firms already face international competition, and businesses are increasingly in need of bilingual individuals to help

> **• FAST FACT •**
>
> In much of the world, language skills are crucial to business success: For instance, among executives, 100 percent in Hong Kong, 97 percent in Singapore, and 95 percent in Indonesia can negotiate in at least two languages.

them stay competitive. According to some experts, the language differential payoff (the higher rate of pay for bilinguals over monolinguals) is as much as 5 to 20 percent. In thinking about your children's futures, it's no sillier to consider your children's linguistic capabilities than their literacy levels or math skills! All three are important for success in today's world.

This trend toward multilingualism is noticeable in all types of schools. The number of elementary schools offering immersion education programs in the United States has dramatically increased in recent years. In many cities, at the most elite and competitive schools, education is often conducted *entirely* in a foreign language. For instance, in the Washington, D.C. area, one of the most competitive schools in terms of admissions is the Washington International School, which offers several tracks, with two of the most popular languages being Spanish and French. At many such schools—private and public—students with some background in an additional language (or languages) are given greater preference in the highly competitive admissions and financial aid process.

This picture is very different from the one that most of today's parents encountered as teenagers, when two years of introductory high school French or Spanish were considered standard and sufficient. Within our increasingly globalized world, our children will need a second language much earlier and at a much higher level of competence than we could have imagined twenty years ago. Given this need, it makes sense to give our children this educational and career advantage by starting young.

Clearly, there are many reasons why parents might want their child to speak more than one language. Before we delve into the nuts and bolts of how, when, and where, we've found that it is useful for parents to clarify (both within themselves and sometimes *between* themselves) the why component. Take a few minutes to consider why you want your child to be billingual. (See the Exercise on page 265.)

Now that you are aware of many of the advantages that come with knowing more than one language, let's get down to the nitty-gritty of *how* you can best achieve this goal. We'll start by wiping the slate clean and debunking the top ten myths about how languages are learned (and how they aren't) in the next chapter.

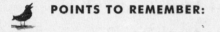 **POINTS TO REMEMBER:**

- Children who know two languages can gain a cognitive, academic, and social edge over monolinguals.
- Learning another language may promote positive cross-cultural attitudes, behaviors, and friendships.
- Knowledge of one's heritage language can promote greater self-esteem and self-confidence.
- Advanced knowledge of more than one language is crucial in today's multilingual, multicultural world.

CHAPTER 2

Myths and Misconceptions about Learning a Second Language: What's the Real Deal?

All parents wonder and worry if they are doing *it* right. From the very beginning, we are bombarded with questions and choices (natural or medical birth, breast-feed or formula, and so on). For most of these decisions, we rely on our intuition and the wisdom and advice of friends and family. We sometimes consult doctors or books. However, when it comes to making decisions about language, today's parents are often left hanging.

With language, as with any other areas of child-rearing, parents have many concerns and questions: "How can I best help my child to learn a second language?"; "Can my child learn a second language from TV?"; "Will she be a late talker if she's exposed to two languages?"; "What if he starts mixing up languages?" Educators often have some of the very same questions. Figuring out the answers is not always easy. For starters, there's very little information on language in any of the standard parenting books and what is written in these books is often not based on scientific findings. Moreover, getting to the bottom of these issues is complicated by the fact that, in a sense, *everybody* is an expert on language learning. After all, we've all learned at least one language and many of us have watched our children or the children of friends and family learn to talk in those first few magical years. As a result, we have shared commonsense ideas about language and language learning.

However, sometimes these shared understandings are *not* based on any scientific evidence. A good example of this is the very widespread notion that children learning two languages will experience language delay. Contrary to popular belief, this idea is *not* a conclusive scientific finding. Actually, it is a prime example of a language learning myth. By "language learning myth," we mean a piece of popular wisdom that well-intentioned friends, grandparents, parents, neighbors, and sometimes even pediatricians and teachers repeat over and over. Often we hear it so much that we come to believe it and use it as a way of describing and making sense of the things we see around us. But as we will show you throughout this book, popular, widespread, commonsense beliefs about language learning are completely wrong in some cases, and in other cases, they are only partially true.

In recent years, there have been major research advances in linguistics, psychology, and education, and these have led to new insights into how children learn their first languages, as well as how they can learn second languages effectively and efficiently. While this research is fascinating in its own right, it also has important implications for parents who want their children to know more than one language. Indeed, understanding these latest scientific developments—and in particular how they contradict many popular, commonsense ideas about language and language learning—is critical to ensuring optimal second language learning outcomes for children. Not knowing how languages are learned (and not learned) can mean that parents and children needlessly waste time, effort, energy, and often money as well. In this chapter, we'll talk about what we call the top ten myths concerning second language learning. We want to address these right at the beginning of the book so we can start with a clean slate to address the real how and when of second language learning in the following chapters.

• FAST FACTS •

THE TOP TEN MYTHS ABOUT SECOND LANGUAGE LEARNING

1. Only bilingual parents can raise bilingual children (and bilingual parents always raise bilingual children).
2. I'm too late! You have to start very early for second language learning, or you will miss the boat.
3. Only native speakers and teachers can teach children second languages.
4. Children who are raised in the same family will have the same language skills as one another.
5. It's important to correct errors as soon as they appear in grammar and vocabulary (to prevent the formation of bad habits).
6. Exposing my child to two languages means she will be a late talker.
7. Mixing languages is a sign of confusion, and languages must stay separate (one-parent–one-language parenting is the best way).
8. Television, DVDs, and edutainment, like talking toys, are great ways to pick up some languages.
9. Bilingual education programs are for non-English speakers.
10. Two languages are the most to which a very young child should be exposed.

Myth #1: Only bilingual parents can raise bilingual children (and bilingual parents always raise bilingual children).

Many of us assume that bilingual parents raise bilingual children and monolingual parents bring up monolingual children. End of story. While this seems logical enough at first glance, in fact, it's not true. Let's take the two points in order. Decades of research and mountains of data tell us that bilingual parents most definitely *do not* always raise bilingual children. The history of immigration within the United States tells us this story as well. Immigrants arrive often speaking only the language of their native country (Spanish, Tagalog, Farsi, etc.); the children of these immigrants are generally bilingual in that heritage language and in English; and the grandchildren of these immigrants are most often monolingual in English. This switch to English monolingualism happens in just a few generations. If bilingual parents always raised bilingual children we'd have many hundreds of thousands of U.S. citizens who also speak German, Italian, Mandarin, Polish, and many, many other languages!

The truth is that raising bilingual children takes planning, effort, and dedication, even for parents who are bilingual themselves. There are (at least) three reasons why this is so. First, in the United States, monolingualism is considered the norm, and child rearing in one language is seen as the most typical. Second, English is a high-status, high-prestige language. For children in the United States (and beyond), this means it's the language associated with nearly everything fun and cool. And third, wherever they live, children—even very young ones—are aware of the status encoded in language. This means they pick up very quickly on who speaks which language when (and for instance, that most of the other big boys that they admire in the sandbox speak English, *not* Korean or Portuguese). Taken together, this means that bilingual parents are swimming against a very powerful monolingual riptide! In order to be successful in raising bilingual children, parents need to be prepared with the best and most current research findings on second language learning as well as many effective and fun tips and tricks based on this research. (Both of which we'll get to in the pages that follow.)

Monolingual parents face many of the same challenges as bilingual parents, but with the added burden of not being able to use the second language fluently themselves with their children. However, for families like these, in many ways, things have never looked better. There are now literally thousands of opportunities and hundreds of ways for children to learn second languages in the United States. Bilingualism is increasingly seen as a highly desirable asset. And as a result, there are now more resources than ever, ranging from classes to bilingual toys like Dora the Explorer to the iPod (more on this in chapter 7). However, making sense of these opportunities and separating the good from the useless is not easy. Parents must pick and choose wisely among programs, activities, and materials. Some of these are fantastic and lay a solid foundation for successful language learning, and others are slickly marketed but gimmicky wastes of time (and money!). *However, with the right foundation of knowledge, any parent can raise a child who knows more than one language, even if that parent is monolingual.*

Myth #2: I'm too late! You have to start very early for second language learning or you will miss the boat.

While age clearly plays a role in second language learning, it's simply not true that the only time you can learn a second language is when you are very young. Where did this idea come from? Among researchers this notion is referred to as the Critical Period Hypothesis (and notice it's hypothesis and not fact or finding!). The gist of this hypothesis is that second language acquisition must occur early in childhood and definitely before puberty in order for the child to be a very successful speaker of that language. We'll discuss this in more detail in chapter 4. What we know is that this is not a hard and fast rule because many older children and some adults do achieve very high proficiency even though they began to learn their second language *after* puberty. Indeed, given the right conditions, older children and even adults can learn a second language very well!

For these reasons, some researchers argue that other factors, like motivation and anxiety levels, as well as the amount and type of exposure to the new language that children and adults receive, are critical. For example, consider a Russian family that moves from Moscow to Oklahoma City. Often the children in that family will make faster initial progress learning English than the adults. After nine months, the kids will be chatting away on the playground while their parents still feel tongue-tied. Although this is often explained solely as the result of children naturally being better language learners, if we look closely, we see that there may be another part to the story as well. The children and adults also have very different language learning opportunities. For instance, children often have many more chances to use English outside the home—in school, in the neighborhood, and playing with peers—than their parents. The language children need to use socially is much less complicated (for example, playing tag versus talking about world events). And because children are used to making mistakes at everything, from falling off a bike, to not knowing the words for things, and because many of them are naturally quite talkative, they are usually less anxious about making mistakes in the second language. So, the nature of children's environments mean they are able to dive into second language learning, trying out the new forms right away and often making rapid progress.

In thinking about whether younger is always better, it's also important to separate *speed of learning* and the *ultimate level of success*. In terms of speed, adults and older children are often initially faster at learning some aspects of a second language, including complicated grammar. In terms of ultimate level of success, younger children can usually catch up and typically reach higher levels of fluency. In particular, younger learners appear to have a clear advantage in their accents. Yet even some older children and adults, in the right environment, can achieve extremely strong mastery of second language grammar. So, in short, although younger learners have advantages in specific areas, the window for learning a second language never completely closes. *In other words, there are some advantages to starting very young, but older children can also make great strides and reach high levels of success in the second language.*

Myth #3: Only native speakers and teachers can teach children second languages.

All parents worry about modeling good behavior. In terms of language, for some parents this means making an extra effort to use "please" and "thank you" around their children, or *not* to use particularly strong words. Other parents worry that their children should be exposed to native language models, especially if the children are learning more than one language. They are right that adults and other speakers play a very important role in children's language learning. However, they are wrong to think that children will learn only if the adults provide them with perfect linguistic input and exposure. This holds true for both first and second language learning.

One of the most amazing things about a child's first language learning is that it happens naturally and flawlessly despite the lack of perfect speech that surrounds (and is directed at) children. All children end up knowing how to speak much like the adults around them even though much adult speech (in *any* language) contains false starts, hesitations, interruptions, backtracking, sentence fragments, and grammatical errors. (Indeed, when researchers analyze recordings of adult native speech, we are often hard-pressed to find many full and grammatically complete sentences.)

If someone is speaking her second language (and is not a highly proficient or nativelike speaker), then she may say simplified and sometimes ungrammatical sentences. However, these sorts of imperfections do not harm or impede children's language learning. In other words, for both first and second language acquisition, children get enough to learn (and eventually master) the language. Research shows that as long as children have experiences with adults or older siblings interacting through language, they develop the ability to use it. So, is it critical to have a native language model who speaks in complete sentences? Children seem to be responsive to language that is tailored to their developmental levels, complete sentences or not. What is critical is not that children hear complete sentences but that they are directly engaged in conversation. Children have an amazing ability to learn language, and acquisition occurs even though adults do not always speak perfectly or do not actively teach them a language. Directing sophisticated language to the child from the outset is not crucial. Even parents with limited second language proficiency can interact with their child in the second language, providing important language input. The value lies in the interaction. This is not to say that non-native-speaking parents should not supplement their own input by finding other opportunities for their children to interact with native speakers. However, it's a myth to assume you need to be a native speaker to provide quality second language interaction for your child. Parents who have some limited second language skills can still provide an important foundation in the language. *The truly critical factor is rich, dynamic, and meaningful interaction with speakers of those languages (and this can come in many different forms).*

Myth #4: Children who are raised in the same family will have the same language skills as one another.

Brothers and sisters often share a lot in common. They may all have their mother's smile or their father's slightly goofy laugh. Only very rarely, however, do siblings have the same language proficiencies. This is for two sets of reasons. First, just as any parent of more than one child will tell you, there are always going to be some innate differences between two siblings, even between twins, and even if raised under very similar circumstances. As we detail in chapter 5, subtle differences in personality and memory skills, for instance, can impact language learning.

Second, even within the same family, two children can have very different experiences with language learning. For instance, researchers who have looked at language patterns in U.S. homes where parents speak (and want their children to learn) a non-English language, such as Chinese or Spanish, have found big differences across siblings. First-born children are *far more likely* to speak their parents' language than second- or third-born children. This makes sense, of course. Parents naturally have the most one-on-one language and interaction time with their first-born child. Subsequent children, in turn, spend much more time in conversations not just with their parents, but also with their older siblings, who often introduce English into the mix. Later in this book, we provide some tips to help balance things out. For now, *keep in mind that for lots of different reasons, children reared within the same home can end up with very different language skills.*

Myth #5: It's important to correct errors as soon as they appear in grammar and vocabulary (to prevent forming bad habits).

Errors are a natural and expected part of language learning. This is obvious to parents because all children make mistakes in learning to talk. Parents naturally ignore many if not most of their young children's mistakes, often finding them cute and repeating them ("piwano" for *piano*, "busketi" for *spaghetti*, or "crocodidle" for *crocodile*, for example). Parents instinctively know that they cannot and should not correct each and every error their children make. Their children's confidence might suffer. They might be less motivated to talk. And both parents and children instinctively know they should mainly focus on communication and understanding, not the grammatical accuracy of what the children are saying. In the same way, if a teacher tries to correct all language errors in a classroom, or a caretaker or grandparent constantly corrects a child speaking in a second language, it is unlikely that the child will respond positively by continuing to try out new constructions, or by enjoying playing with the language to learn.

Today, most researchers and teachers agree that error correction should be done selectively, and in ways that help learners to notice and discover what is needed for their second language development without their motivation and confidence being negatively affected. When learners make errors, this is an opportunity for teaching, but one to be used with caution. And in fact, sometimes making mistakes is actually a sign of progress.

We often assume that mistakes occur in the speech of second language learners as a result of their problems with the second language, usually because of some kind of interference from the first language. In other words, if your first language is Mandarin Chinese, which does not have definite (like "the") and indefinite ("a") articles, you may have some problems with "the" and "a" in English and leave them out. However, researchers who have examined speech samples from second language learners discovered that, in fact, *most* of the mistakes learners make are *not* due to differences between their first and second languages.

From looking closely at the speech of many, many learners, researchers have observed similar patterns in the ways that learners make mistakes. They found stages or routes of development that all learners go through when acquiring a second language, *regardless of their first language.* For example, most learners of English begin asking questions by putting a rising (question-style) intonation onto a sentence: "You go to store?" This first step, of course, is not standard English. As all learners go through this first step (and later ones as well) in the same sequence or basic order, we assume that a shared mental mechanism (and not their first language) is driving their language learning process. This means that mistakes need not always be viewed as problematic. Learners become proficient because they are given as many opportunities as they need to use the target language, and because some of their errors are creatively corrected, not because every error is corrected each time it occurs. In other words, then, *errors should be treated with care and often ignored, since the value of communication is paramount in both first and second language learning.*

Myth #6: Exposing my child to two languages means she will be a late talker.

One of the most common misconceptions about early language learning is that it will result in language delay, in other words, that children will talk later or less than they would have if spoken to in just one language. Why is this belief so prevalent? Probably because there is a great deal of variation in the ages at which children begin to speak. For instance, research with thousands of children tells us it is normal for a child to begin to utter her first words as early as eight months or as late as sixteen months of age. So in any given group of one- and two-year-olds, there will very likely be children who are chattering away, children who are producing simple one- or two-word utterances, and children who are still pointing and grunting. Unfortunately, if the child who is learning two languages is among the point-and-grunt group, there is a (natural, but misguided) tendency to blame the child's additional language.

Yet as we'll discuss in detail in chapter 10, there is no scientific evidence to show that hearing two, three, or more languages leads to delays or disorders in language acquisition. Monolingual and bilingual children begin to babble, to say their first words, and to speak their first two- and three-word mini-sentences at about the same time. Indeed, many, many children throughout the world grow up with two or more languages from infancy without show-ing any signs of language delays or disorder. This is true across different languages, for young children and for older children. So parents can cross language delay off their list of worries right from the start. *Learning two languages is not a cause of language delay.*

Myth #7: Mixing languages is a sign of confusion, and languages must stay separate (one-parent–one-language parenting is the best way).

Many of the parents we have worked with have also expressed con-cern that their child might become confused by the use of two languages. The main worry is that young children might not be aware of the presence of two different language systems or able to understand that two different words can refer to the same concept. Like parents, researchers have also wondered about whether chil-dren are confused about interacting in two languages.

Results of decades of carefully conducted research point to the fact that young children distinguish early on between their two languages. This form of linguistic practice or mental exercise has even been linked to greater scores on certain intelligence measures. Mixing languages is a normal phase of bilingual language develop-ment. It seems to be near universal among bilingual children and is apparent even at the babbling stage; that is, long before children can say a word in any language. While the precise cause and func-tion of mixing is not known, the end result is very clear: *All children move beyond this phase.* While often unsettling to parents, the mixing phase is generally short-lived, finishing long before formal school-ing begins, and is not problematic in any way in the long run.

Many studies have shown that children are very sensitive to the unspoken rules about which language should be spoken to whom and when, and naturally sort this out on their own. Children do this *without* any explicit help or teaching from parents or teachers. As children learn language (for example, vocabulary words, grammatical rules such as -*s* at the end of "bear" means more than one bear), they also learn how languages are used socially in interactions (for example, that it is okay or not okay to interrupt, or that throwing dinner on the floor will get your parents to stop talking about the credit card bill but might also attract some attention of the unwelcome sort!). In addition, they learn which languages should be used with whom. So while some parents believe that strict separation of languages by person is the only or best way to raise a child bilingually, this is definitely not the case. There are lots of documented cases of children who grow up bilingually hearing both languages spoken by both parents, as well as children from one-parent–one-language homes who understand two languages very well, but only speak one. As we'll discuss further in chapter 9, parents should focus on the quality and quantity of input children receive in each language and not worry about maintaining strict separation. So in short, the "real deal" here is: *Most children go through a period of language mixing. It's normal! There are many different ways of organizing language in the home, and strict separation of languages is generally not realistic and not necessary.*

Myth #8: Television, DVDs, and edutainment, like talking toys, are great ways to pick up some languages.

Parents today have a wide variety of language-teaching materials to choose from, ranging from tapes, DVDs, and computer programs to trilingual talking toys. In our first months of motherhood, we were delighted to pop in "edutainment" DVDs like the *Baby Einstein Language Nursery Video* or a little something from the Muzzy series. To our minds, not only were we exposing our children to valuable foreign language input, possibly making them brainier (according to the advertising), but we also got a few moments to check our e-mail and have a hot shower. However, after a while, we started wondering: What types of language learning can take place through simply watching these popular videos? And how should parents balance any language benefits with the fact that the American Pediatric Academy recommends *no television at all* for children under two? Or the fact that a new commercial-free pay TV channel, BabyFirstTV, was launched in 2006, amid controversy and criticism from the medical profession, who point out that while the channel's creators say that it will help parents and babies bond, TV seems to be a very weird thing to place in the bonding equation. This leads us to crucial questions: Is one type of exposure better than another? What sorts of language learning can take place through television?

Research with very young children tells us that even small amounts of foreign language exposure can help children keep an ear for that language. In other words, just an hour a week of interaction in the language can help a child still hear tiny distinctions in the sound system of a language, much like a native speaker of that language. But—and here's the rub—to be effective, this exposure must be with a real human being, and not a DVD, television program, computer game, or talking toy. These edutainment devices, while extremely well marketed and popular with parents and some children, cannot substitute for a real person and real interaction. However, that living, breathing human being does *not* need to be a credentialed language teacher, nor does the person need to be a native speaker of the language. In other words, parents with even minimal skills in a second language can interact in that language with their young children, for instance, by reading simple bilingual books or playing basic games like peek-a-boo in the foreign language. These sorts of activities do much more to promote language learning, and other types of learning as well, than baby videos of any sort.

Parents of older, school-age children can feel more comfortable about using DVDs and the like to support their language learning goals for their children. Used judiciously, television in the foreign language can provide a positive and fun association with the language and help meaningfully link the language with culture. And older children, especially those who can read the English subtitles or who know a bit of the language already, do seem to learn some language through these fun and enjoyable programs. We'll discuss more of the research on the pros and cons, together with specific tips in chapter 7. For now, parents should keep in mind the simple point that *children don't learn much language through television or other edutainment products—these should be thought of as supplemental for older children.*

Myth #9: Bilingual education programs are for non-English speakers.

Many parents mistakenly think of bilingual education as designed to help immigrant children learn English. While this was true in the past, the terrain of bilingual education has radically shifted in the last decade. On the one hand, state initiatives in places like California and Arizona—together with federal education policies like No Child Left Behind—have meant that traditional bilingual education programs have been dramatically scaled back in many areas. In those that remain, English language learners typically receive large doses of English as a Second Language classes and small amounts of instruction in their native language for a short period of time. These programs tend to transition children to English dominance or English monolingualism quickly. However, it's not all bad news. A new type of bilingual education, known in education jargon as dual language immersion or two-way immersion, has soared in popularity in recent years. In the United States, there are now more than 350 of these dual-language schools; in Canada, there are also hundreds of immersion schools and programs in operation.

What is dual language immersion and who is it for? These schools and programs (as well as a growing number of language-immersion summer camps) aim to provide academically rigorous instruction in two languages to two groups of students: native

English speakers and native speakers of another language (for example, Spanish). The goal is for both groups of students to become bilingual and biliterate after a few years in the program. In a typical dual-language school, half of the children will be speakers of one language (for example, Chinese) and half are native English speakers. Instruction is provided through both languages. The research results from these schools are very impressive so far. Overall, studies in these schools show that both groups of students achieve high levels of competence in both languages and work at or above grade-level in academic areas, such as mathematics and reading. Not surprisingly, demand for such programs is booming as well. In chapter 8, we discuss how interested parents can identify and evaluate the programs available in their area. For now, the point to take away is: *Bilingual education programs can be beneficial for all children.*

Myth #10: Two languages are the most a very young child should be exposed to.

So learning two languages sounds good, but what about three? For many parents, concerns about language learning—language confusion, delay, and mixing—are intensified when the decisions revolve around three languages instead of just two. So for instance, can or should a family who speak Italian and English at home enroll their children in a Spanish-English dual immersion program like the one described above? Although we don't have much research yet on *how* bilingual language learning differs from trilingual language learning, we have a lot of successful and healthy human data walking around the world. In other words, millions of children grow up learning three or more languages and have been doing so for hundreds of years, with *no ill effects.* In many parts of the world, in Singapore and Switzerland, for instance, it is common for a child to learn one language at home, another in her community, and a third at school. Given the right conditions, these children can learn these languages well enough to be able to communicate competently and appropriately in each context.

However, trilingually oriented parents (like bilingually oriented ones) do need to take special care to make sure that the child has adequate opportunity to engage in rich and meaningful interaction in each of the languages regularly. In other words, the child needs rich *quality* and intensive *quantity* of language. These parents also need an understanding of how languages are learned and a large bag of tips and tricks to ensure fun, active, and interactive language learning for their kids. Nearly all of the points we discuss in this book concerning bilingualism equally apply to promoting trilingualism. For now, parents should rest assured that *the more the merrier—three languages (or more) are possible.*

WRAP UP

Much of what we commonly believe about language learning has been scientifically tested in recent years. Some of it has turned out to be based on myths. These myths can be harmful because they point us down the wrong paths, preventing us from making the most effective decisions for our families. As we hope we've made clear here, the real deal on these top ten myths differs sharply from popular wisdom.

Now that you know both the real scoop on how languages are learned *and* the undeniable advantages of bilingualism, you're ready to begin planning your strategy for your child's language learning. In the chapters that follow, we'll guide you through the basics of *when* and *how* to best achieve this goal. First, however, let's turn to the question of *which* language is right for your family.

• FAST FACTS •

THE REAL DEAL ON THE TOP TEN MYTHS ABOUT SECOND LANGUAGE LEARNING

1. Any parent can raise a child who knows more than one language, even if that parent is monolingual (all children can learn a second language even if their parents don't know that language).
2. It is never too late and younger is not always better in every way.
3. Rich, dynamic, and meaningful interaction is critical and more important than having a perfect native-speaker model.
4. For lots of different reasons, children reared within the same home can end up with very different language skills from one another.
5. Constantly correcting errors can do more harm than good.
6. Learning two languages is *not* a cause of language delay.
7. Most children go through a period of language mixing (it's normal!). Strict separation of languages is generally not realistic and not necessary.
8. Children don't learn much language through television or other edutainment items—these should be thought of as supplemental.
9. Bilingual education programs can be beneficial for *all* children.
10. The more the merrier—learning three languages (or more) is possible!

Which Language and When?

CHAPTER 3

Which Language Is Right
for Your Child?

While bilingualism might sound ideal, if you're like a lot of parents, you're quite possibly scratching your head and wondering: "Okay, but *which* language?" For some families, the question of which language is pretty easy and obvious. Kendall, for example, is a fluent second language speaker of Spanish and always knew she wanted to bring up her child as a Spanish-English bilingual. Alison, on the other hand, has a low level of competency in a couple of different languages, as does her husband. She's not the multilingual kind of linguist, but rather the hopeless-at-learning-languages kind. Choosing a language can be a tricky task for some families while being a straightforward decision for others.

For example, Becky, English-speaking mother of two-year-old Ben, always knew that she wanted Ben to learn German early on. She had struggled to learn German through grammar-focused foreign language classes in high school and college. Becky only really got the hang of the language after living in Germany for several years while working as a business consultant. She experienced first-hand the edge this provided her in her work, but also in expanding her social and cultural life in Germany. Becky really wanted Ben to have the same sorts of advantages. German made sense for her family because it was the language she knew best (other than English of course).

For other parents, the issue of which language is not so clear. Mika, for instance, grew up speaking Japanese with her parents (who had come to the United States from Japan as young adults). By the time Mika was a teenager, her parents spoke a mix of Japanese and English with her. She can understand Japanese but doesn't feel very comfortable speaking it. She and her American-born boyfriend, James, both studied Spanish for a few years in high school, but are far from fluent. She's eager for her four-year-old son, Jacob, and her new baby, Katie, to learn a second language. But she's torn about which one and how to decide: Japanese? Spanish? Or perhaps a different language altogether, like Chinese?

FIRST LANGUAGE LEARNING:
BABIES ARE CITIZENS OF THE WORLD

Before we delve into how to pick which additional language is best for your child, it's important to cover some of the nuts and bolts of first language learning. All babies come into the world with a gift for languages. From the day they are born, they begin to learn the language that surrounds them. In this way, every baby is a citizen of the world with full and equal capacity to learn any of the planet's six thousand languages.

During the first few months of life, all normally hearing newborns are able to distinguish between an incredibly wide range of sounds in many, many different languages. (This includes sounds that adult speakers of certain languages can no longer hear—for instance, monolingual speakers of Japanese don't hear the difference between *l* and *r* so words like *lake* and *rake* sound very similar to their ears.) Researchers have tested infants' skills by exposing them to similar sounds (for example, *la* and *ra*) from different languages and then measuring how hard they suck a special pacifier that records sucking strength. In these experiments, babies suck pacifiers harder and faster when they hear something new, and when they are a bit older, they turn their heads for the same reason. From this type of research we know that even if, for instance, Japanese infants have never heard a word of English, they are able to distinguish equally well between sounds like *la* and *ra* in English. And, for example, Italian babies who have never left their own neighborhoods can hear differences between various click sounds in African languages such as Xhosa. This makes sense. Human babies are equipped with the ability to pick up any language equally well. Infants and children learn the language (or languages) that surrounds them and is used with them in everyday life.

So, let's say your baby girl is being raised in an English-speaking environment. What happens when you expose her to sounds that most adult native speakers of English *can't* tell apart? For example, in Hindi, there are two sounds that both resemble *t* in English. In Hindi, these two sounds are produced in slightly different parts of the mouth. If you say one *t* sound instead of the other, you might as well be saying a different word entirely (like the difference between *lake* and *rake* in English, in which the only difference is *l* versus *r*). Can your baby hear the difference between such similar sounds even though she's never been exposed to Hindi? It turns out that she can—but only until the ripe old age of about ten months. Even before her first birthday, this ability to hear the difference between sounds that are not found in the native language begins to decline.

This makes good sense: Babies don't know where they are going to live and what language they will need to speak, so they need to be born with the ability to hear the difference between all sounds. The timing of the loss of this ability also makes sense: Infants' skill in hearing these differences declines just as they are producing their first magical words. As infants tune in to the main language(s) around them, they begin to tune out the sorts of distinctions that are not important for their language(s), but are important in others.

ASSESSING LANGUAGE IN YOUR FAMILY

As this tuning-in process happens fast and each family situation is unique, when deciding the question of which language is best, you should start by considering the language profile of your family. Given the importance of interaction in the language discussed above, this is an essential step. The following questions can help you get started on this process. We recommend that you spend a few minutes considering these questions before moving ahead. For the best results, consider them with the other family members or caregivers who have a role in raising your child.

 EXERCISE:

Assessing Your Family's Own Language Profile

1. What languages do you know and how well do you know them?

_____

2. How comfortable do you feel using each of these for: Singing lullabies? Reading bedtime stories? Explaining why the sky is blue and why cows eat grass? _____

_____

3. If you have a second or third language, what would you *not* feel comfortable doing in that language? _____

_____

4. Who else in the family speaks each of these languages? _____

_____

5. How comfortable do they feel using those for: Singing lullabies? Reading bedtime stories? Explaining why the sky is blue and why cows eat grass? _____

_____

6. What would they *not* feel comfortable doing in those languages?

_____

7. How often do you use each language in everyday life at present (with friends, on the phone, with family members, at work)? _____

_____

(Note: Don't worry if you only have one language in your family. We'll get to that soon in this chapter!)

There are no right or wrong answers to these questions. The point is to start thinking (and talking!) about your family's own language attitudes, skills, and resources. Language is very intimate in lots of different ways. Discovering how you feel about using certain languages (or how your loved ones feel) often requires a bit of emotional digging. Our experience tells that investing some time in this sort of digging at the outset is well worth it in the long run.

We struggled with some of these issues ourselves. In fact, these unexpected challenges were part of the motivation for writing this book. For instance, as Kendall is a proficient speaker of Spanish she has no trouble making conversation in Spanish with adults. She thought a lot about child bilingualism while pregnant, but found herself at a loss for words when Graham finally appeared. She felt silly and awkward using Spanish with him at first. Although she had used Spanish professionally for years, she hadn't been exposed to that much baby talk in Spanish. While speaking Spanish with colleagues and adult friends was easy, Spanish lullabies were not. At first, speaking Spanish with Graham felt artificial and all wrong! She needed a bit of time to get into a new way of speaking.

Even native speakers can be caught unprepared in this way. For instance, a Korean-American friend, Sook, was planning on speaking Korean to her twin boys. Sook is fluent in both Korean and English, having grown up in Korea and moved to the United States at seventeen. However, once her boys were back from the hospital, she found herself struggling. As she explained one day, "When they were really small, I realized that I didn't want to raise them in that strict Korean way. It brought up a lot of emotion for me. I felt dominated at times by my parents and grandparents. In the Korean language, there are lots of markers of respect and hierarchy built into every sentence. I worried that by using only Korean, I'd be forcing them to participate in that old-fashioned and formal system. I wanted a more American and less formal relationship with my kids and I had to figure out how to do that in Korean."

Through some trial and error—and mostly by getting over feeling self-conscious about using language with their children in ways that were different from how *their* mothers had talked with them as children—Kendall and Sook managed to cope with these surprise emotions. They both went on to use mostly Spanish or mostly Korean with their children. The point is, however, that they could have been better prepared. One of the best ways to prepare yourself and to avoid such surprises is to carefully consider your family's language profile. That can mean the language profile or biography of the extended family, not just the parents. Sylvia, for example, is a mother who participated in one of our past studies. Sylvia's first language is English. She knew from the start that she'd be raising her child on her own and wanted her soon-to-arrive baby girl to be bilingual in Spanish and English. The father of the baby spoke Spanish, but wasn't going to be regularly in the picture. Sylvia spoke some Spanish, but not a lot. As she thought about her own language profile, she realized that she would be comfortable singing little songs and reading Spanish books to her daughter. Her ex's parents were eager to be a part of their granddaughter's life. And although they were not fluent readers, they could provide lots of rich conversational stimulation and interaction each week. Furthermore, since she lived in a multicultural area on the edge of a big city, there were quite a few different Spanish-based children's activities nearby. Sylvia realized that it was realistic and feasible to aim for a bilingual baby—once she carefully pieced together all of the resources of her extended network.

What about families where only one parent speaks the second language? How do they all deal with the issues involved? Obviously, they have made a decision about which language, but as the child's proficiency improves, one parent may begin to feel left out. The other parent might begin to feel as though they bear a disproportionate burden in terms of providing language exposure and interaction. Our friends from graduate school, Rita and Anthony, live in Singapore, where Anthony was born. Rita was born in the United States and speaks English (fluently), French (some), and Japanese (a little). Anthony speaks English fluently, and also Mandarin, with a mid-level conversational ability. They have a daughter, Emily, who they adopted from China. They hope to raise her to be bilingual. Mandarin is widely spoken in their local area and Emily goes to a bilingual Mandarin-English day-care center. Emily at age three and a half already expresses a strong preference for English and usually only speaks Chinese to her Chinese-speaking day-care teacher. Without much support for the language at home, it seems likely that Emily's Chinese will develop into a school language rather than the rich resource her parents want it to be. However, at home when Anthony occasionally says something to Emily in Chinese and Emily answers, Rita is left to ask, "What are you saying?" Rita wonders if she'll feel left out as Emily's proficiency develops. If they are still living in Singapore when Emily is a teenager, will she speak Chinese with her friends when she doesn't want her mother to know what she's saying?

Hopefully, in situations like this, both parents will be able to collaborate in locating materials, programs, tutors, babysitters, summer camps, and all the other kinds of language support that we discuss in this book. And hopefully the parent (and eventually the child) who speaks the second language will be sensitive to the one who doesn't. In some families we know, the left out parent has made an effort to learn the language as well so as not to be excluded. However, it is important when selecting which language to think about the long-term from the outset and weigh all of the factors as carefully as possible.

So, as you start to answer the question of which language, the first place to start is by taking a good hard look at the language resources and language attitudes within your own immediate and extended family. You may find some surprises! Clarifying *who* can do *what* with *which* language (and how they feel about doing it!) is a crucial first step.

ASSESSING LANGUAGES IN YOUR COMMUNITY

If you don't have a special connection to any particular language, then big-picture issues will probably be important as you decide the which language question. As we mentioned in the introduction to this chapter, Alison is not a bilingual. In fact, she credits her interest in second language learning research to the difficult time she has always had learning foreign languages. When she was ten, her family moved to Wales from England. She attended a primary school where the Welsh lessons were conducted only in Welsh. If Alison spoke a word of English by mistake, she had to write it on a piece of paper, go into the school yard, dig a hole, and bury the piece of paper. Alison was outside burying paper quite often, and it rains a lot in Wales! Hence her interest in how some people (adults and children) learn—or don't learn—languages.

Today, Alison knows a bit of French, Japanese, and, of course, Welsh, and her husband knows a bit of Spanish, Nepali, and Japanese. Since Japanese is the language they have in common, they decided they would work to teach their daughter Miranda some Japanese. However, because they also have Spanish resources in their community, they decided to also expose her to Spanish. When Miranda gets older, if she prefers one over another, they really don't mind. The benefits of knowing a second language apply regardless of which language is learned!

Along these lines, it's helpful to consider Jim's situation. Jim always thought he'd teach his children French from a young age. He had struggled to learn French in high school and only really got the hang of it after living in Paris and working as a bar hand for six months after college. When he found out he was going to be a father, he started getting serious about this idea and doing a little research, much like the exercise below. As he looked (and listened) around his Brooklyn neighborhood, he became more and more aware of the enormous Russian-language resources in his midst. After English, Russian was the number-one language on his block. In his building, there were three separate families with Russian-speaking teenage daughters who were eager to earn a bit of cash from babysitting. The bookstore down the street had loads of Russian children's books and even a bilingual story hour for toddlers. It was the same story at his local branch of the library. There were also three different Russian-English home-based day cares for the parents to choose from right in their neighborhood, and a two-way Russian-English bilingual school was being planned in their district. Jim had never really considered Russian as a language his family might learn, but after comparing the resources in his community for French versus Russian, he decided this made the most sense.

As you consider the pros and cons of different languages, it's important to keep in mind that many of the cognitive, creative, and intellectual advantages of bilingualism apply to bilinguals *in any two languages*. That is the good news! However, this is true *only* if skills in both languages are relatively high. So if there is no obvious second language choice based on your family background or language skills, you'll probably want to identify the language that provides the greatest opportunities and incentives for learning.

This means considering two sets of related issues. The first is which languages are most feasible given your location. Complete the short exercise below to help you identify language opportunities in your area.

 EXERCISE:

Which Language in My Area Provides the Most Language Learning Opportunities?

1. Which languages are spoken in my immediate and surrounding community?

(Hints and clues: You can find the stats at http://factfinder.census.gov/—follow the links to "People" and "Origins and Language" and search for a fact sheet about the languages spoken in your city and state. You can also check your local school district's Web page; districts often post the languages spoken in their schools on their Web sites as well.)

2. Which languages are spoken in my neighborhood?

(Hints and clues: Look and listen closely to what's around you. What languages do you hear in the parks and the grocery stores? What is the language of community signs and newspapers and common announcements? Don't be shy to strike up conversations with speakers of other languages in English.)

3. What sort of language classes and children's activities exist in different languages?

(Hints and clues: Call your community center. Look in local newspapers and parenting newspapers and flyers. Do Web searches. Phone your local YMCA or JCC. Cruise your local bookstores for bilingual readings. Look at notice boards everywhere you go.)

This community assessment exercise will hopefully help you decide which language gives you the best chance of success with your child. For instance, if there are other families who speak that language close by, it will be much easier to form a successful language-based parent and child group. If there are speakers of that language in your neighborhood, you'll have better luck finding a babysitter who can watch your children and teach them a bit of a

---

• FAST FACT •

WHO SPEAKS WHAT IN THE UNITED STATES?

Language	Number of Speakers (in millions)	Percent of U.S. Population (approximate)
English	216	81%
Spanish	32	12%
Other Indo-European (French, German, Russian, and others)	10	4%
Asian and Pacific Island (Chinese, Korean, Tagalog, Vietnamese, and others)	8	3%
Other (Arabic, African languages, Native North American languages)	2	1%

Source: U.S. Census Bureau, 2005 American Community Survey

second language. And if there are lots of other families who speak that language nearby, there's a greater likelihood that your child can participate in language-based activities, like a Spanish music class or a tae kwon do martial arts class in Korean.

Besides the resources and opportunities for language learning in your own community, the other important factor is the incentives for learning different languages. Or put another way, which languages are more in demand or most in need? In particular, which languages are most widely spoken in your country? Most widely spoken internationally? In your region? And which languages seem likely to grow? These abstract questions translate directly into opportunities in everyday life. Language learning is hard work and having a large and growing number of speakers (and situations for speaking) provides a huge incentive for both children and adults.

When we look closely at the language statistics for the United States, it's evident that Spanish is by far the most widely spoken language in the United States after English. What's more, while Spanish is a growing language, it's also a mobile language. While in the past, many Spanish speakers used to live primarily in a handful of urban centers in the United States (Los Angeles, New York City, Chicago, Houston), in the last decade the biggest trend has been the movement of Spanish speakers to more rural and suburban locations and smaller cities such as Durham, North Carolina, or Tarrytown, New York. All this means there are many opportunities and incentives for learning and using Spanish in more and more places—great news for both English-speaking and Spanish-speaking parents! But what about the global picture?

When we look around the world, it's evident that Chinese, and in particular, Mandarin Chinese, far outnumbers the other languages. Not only is Mandarin big today, it's likely to get even bigger. For parents with young children, this means that in ten to fifteen years' time, Chinese could well be the language that offers the most opportunities and the greatest bilingual edge in education and business. Accordingly, the number of Chinese as a second

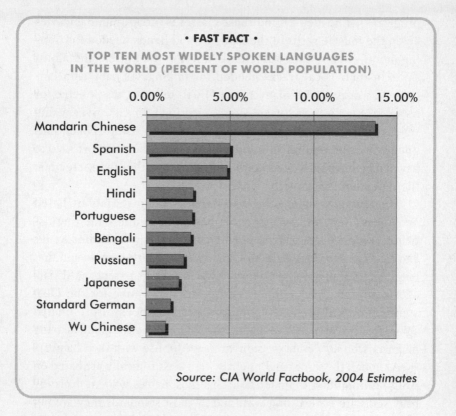

Source: CIA World Factbook, 2004 Estimates

language programs in the United States has shot up sharply in recent years. As some parenting literature puts it, "Chinese could be the new Spanish!" Being at the forefront of this curve by starting young and developing a strong foundation in the language could be a major advantage for your child.

WHAT ABOUT BABY SIGNS AS A SECOND LANGUAGE?

In recent years, baby sign language classes have become very popular (in the United States alone, there are over two hundred programs). The idea behind these classes is intriguing: Babies develop the muscle skills in their fingers and hands needed for signing months before they sufficiently develop the muscle and motor skills in their vocal tracks needed for talking. As baby sign language advocates (and many parents) will tell you, it's possible for babies to engage in two-way communication with caregivers using signs much earlier than they would be able to using spoken language. This no doubt has some advantages. Who wouldn't love to have an eight-month-old sweetly sign "milk" rather than scream at the top of her lungs?

Most (but not all) U.S. baby sign classes and materials are based on American Sign Language (ASL). ASL is a complex and linguistically complete language used by Deaf people in the United States and English-speaking parts of Canada. It is also the native and first language of many people who are born to Deaf parents. ASL is a cherished and rich linguistic and cultural resource for the Deaf community (and ironically, one that has long been discriminated against and treated as less than a full language by the same hearing majority who now embrace sign language for use with their babies).

Although the signs taught in baby sign class typically are based on ASL, it's important to understand that most babies who are learning baby signs are *not* learning ASL and by most standards they are not acquiring a complete language. Through baby sign materials and classes, babies and their caretakers are taught—and often learn to use—a simplified set of signs, typically focusing on basic needs and objects (for example, signs for *more, milk, hungry, thirsty, tired, cookie*, and so on). They do not learn the grammar needed to put these signs together into linguistically complete utterances or sentences. (For example, most people would not say they speak French just because they can count to ten and say their five favorite foods in French!)

This is not to say that baby signs are harmful for babies in any way. Many hearing parents find this an enjoyable and useful way to communicate and even bond with their young hearing children. However, learning baby signs (even when they are based on ASL signs) is not the same as learning ASL, and accordingly, most children (and their caregivers) stop signing once babies learn to speak. Only very rarely (and under special situations, for example, the child spends time with a caregiver who regularly uses ASL) do children continue to develop their sign language skills.

SPOTLIGHT ON RESEARCH:

Do Baby Signs Programs Make Smarter Babies?

To investigate the question of whether baby signs programs help children acquire language earlier or advance their cognitive development, in 2003 researchers J. C. Johnston and colleagues examined seventeen recent studies of baby signing. Johnston found that the studies did not support the big claims made by the infancy industry, that baby signs help make babies smarter or acquire other languages sooner. Many of the studies they scrutinized only looked at a few children, had poor follow-up (they didn't look at their language development for long enough periods of time), and little detail (for example, not enough information on how often the parents and children signed), and overall, the industry's claims went much too far. What's more, the researchers noted that "parents can be stressed by the challenges of meeting demands of work, caring for a young child, and other family and personal obligations, and experience guilt if they feel they are not doing everything recommended by infancy specialists and the infancy industry." The researchers recommended that parents consider their own situations before deciding whether to use baby signs and perhaps consider skipping it altogether. As they note, there are important developments that are taking place before infants begin to speak, such as gaining more muscle control and learning to interact with others nonverbally. Adding an additional task at this important time—*especially* one with no clearly proven benefits—is not essential, the researchers concluded, and children will begin to use language when they are ready.

If you're interested in pursuing baby signs with your young child, or are already enrolled in a program, you might also want to think ahead about language learning opportunities after your child turns one. This might mean continuing to fully develop ASL, if for instance, there are ASL users, programs, and activities in your area. It might mean considering adding another spoken language such as Chinese or Spanish. One thing to keep in mind, though, is that you should not rely on baby signs alone if you want your child to acquire a language other than English.

MAKING THE BEST CHOICE FOR YOUR FAMILY

Deciding which language to teach your children involves weighing the personal, the local, and the global (not necessarily giving equal weight to each depending on your situation). For many English-speaking parents in the United States at least, Spanish becomes the number-one choice. Spanish is sometimes described as the unofficial second language of the United States, as there are millions of speakers residing in all parts of the country. After all, think about how many times you get to "press one for English, press two for Spanish" on automated telephone systems! This means that there are lots of opportunities to form bilingual parent and child play-groups, to create language-focused babysitting coops, or to go to Spanish language cultural events. On top of this, the majority of all bilingual education programs in U.S. cities are Spanish-based (that is, Spanish-English bilingual programs). There is also a great demand for Spanish-English bilinguals in the job market, and so, for instance, even bilingual high school students have an edge searching for a summer job.

Other parents find another language is a better fit for them. For instance, New Yorkers Ari and Diana never thought much about second language learning until their son Sammy turned six months old and Diana needed to think about going back to work. They debated the pros and cons of day care versus nanny. In the end, they felt that the language immersion that a Chinese nanny could provide was worth the additional expense. They decided they would make financial sacrifices in order to hire a

> • **FAST FACT** •
>
> The bilingual edge applies to children who are bilingual in any two languages.

Chinese-speaking nanny. They searched (and paid a premium) for a Chinese-speaking nanny because they think Chinese will be the next hot language. And because New York is home to many Chinese speakers, they easily found the right nanny, and a replacement when the first nanny left, as well as schools and cultural activities to supplement their efforts as Sammy grew older.

Every family and situation is unique and there is no one-size-fits-all solution here. However, there is also good news for undecided parents. As we mentioned above, children can gain the bilingual edge *from any two languages*. It doesn't matter if these language pairings are Spanish-English, Croatian-Taiwanese, Arabic-French, or Portuguese-Japanese. Similar bilingual benefits in terms of cognition accrue for children who attain high levels of competence in *any* two languages. Yet while many advantages have been found for many different language pairings, there is even some evidence that the more different the languages are, the better. For instance, learning Russian, Mandarin, or Greek in combination with English might provide additional benefits, since these languages have entirely different scripts, or writing systems, than English.

WRAP UP

Choosing which language is best for your child is a personal and intimate decision. It's also an important one. Prior to investing time, energy, and effort into language learning, you'll want to consider carefully all of the available resources of your home, neighborhood, and community. Each family and each decision is unique of course. The best decisions (with the best language learning outcomes) result from carefully weighing the personal (how do we feel about using this language?), the local (what opportunities are available?), and the global (what incentives exist?). While there's no one-size-fits-all answer out there for all families, there is a best answer for your family.

 POINTS TO REMEMBER:

- The decision of which language is partly about the family (which language skills do the parents have and how do they feel about the language?) and partly about the broader community (which languages present the most opportunity and incentive?).
- Most baby signs programs focus on sign vocabulary but do not include sign grammar. Unless your child is being exposed to an established, natural sign language such as ASL, learning baby signs is not the same thing as learning a second language.
- Language choice is an intimate and deeply personal one. Parents should consider their own feelings carefully and discuss these issues frankly with all those involved in caring for their child as well.
- What's important is not so much *which* language, but *sticking with it* as very few benefits come with only knowing a few words of a language.

CHAPTER 4

When Should Your Child Start Learning Another Language?

The question of when it's best for your child to start learning another language—at birth, early in childhood, or later on in childhood—can be a tough one. There is a lot of information (and misinformation) out there on this topic, and questions about age and language learning will vary a lot depending on your particular family circumstances. One popular myth that we mentioned in chapter 2 is that "younger is always better," that young children soak up new languages like thirsty little sponges and older children's language-soaking abilities are more like chunks of petrified wood. While it's certainly true that when children learn a second language during early childhood they often end up with much better accents than kids who begin learning a language in high school or college, young children do not have an advantage over older children in every way.

We strongly believe (based on lots of research evidence) that it's never too late or too early to learn another language. *Now* is the right time for your children to learn a second language, no matter what age they are. However, parents should be aware that some things will be a bit harder or easier for children at different ages. In this chapter we provide you with an understanding of language learning abilities and opportunities at different ages so that you can use these insights to your children's advantage. Children at different ages approach language learning in their own ways for a variety of reasons, including their personalities, aptitudes, identities, and levels of social and emotional development. Many of these things are outside of parental control. But being aware of these different factors can also help you make decisions early on, set realistic goals, and then encourage and motivate your children in sensitive, individualized ways throughout the process of learning another language. So we encourage you to consider this chapter on *when* to teach your child a second language along with the next chapter on *individual differences* because in a lot of cases the *when* question will play out differently depending on the particular child. Let's start with some essential background information on how child language learners differ from adult learners socially, emotionally, and cognitively, then we'll turn to how you can leverage your child's age for the best outcomes.

DO CHILDREN LEARN SECOND LANGUAGES DIFFERENTLY FROM THE WAY ADULTS DO?

Social and Emotional Differences

Children are generally spoken to in different ways than adults, whether in first or second languages. For example, compare the language children hear in a bedtime story or in the sandbox with the language adults might hear at a work meeting or read in a newspaper. You'll see there is quite a bit of variation in both language density and complexity! These differences mean that young children hear language that is much easier to process, with many more clues about the meaning of what's being said (for example, the pictures in the book and the actions with the sand toys). Plus, what children need to *do* with language often requires different levels of competence than the complex and often subtle messages adults need to understand. (For instance, handing over a toy truck in response to "That's

> • **FAST FACT** •
>
> Learning a language from a young age is advantageous because older children, adolescents, and adults have to deal with higher expectations and more sophisticated social situations while learning a second language.

my truck!" in the sandbox versus responding to a proposal presented at a work meeting.) You might say that when children start learning a second language early, they find themselves in situations that provide more support (for instance, in the form of visual cues and clues about meaning) for language understanding and use.

Also, young children usually don't face the same kinds of emotional pressure adults have in foreign language learning. They don't have careers riding on their ability to speak a foreign language, and most young children don't take their mistakes too personally and are not embarrassed that they sound silly. Adults often worry about all of these things, but children are used to making mistakes with language. The expectations for adults (both their own and those of others) tend to be greater. These kinds of social, environmental, and emotional factors have a big impact on learning and can help us understand how learning a language as a young child as opposed to even an adolescent has its advantages.

Brain-Based Differences: Critical and Sensitive Periods?

There are also brain-based differences between adults and children, which seem to have led to this "younger is always better" idea so prevalent among parents. In the 1960s and '70s, there was a lot of talk about how there might be a "critical period" for language learning. Critical periods had been observed in other parts of the animal

· FAST FACT ·

Children's second language learning ability changes throughout childhood, but does not completely disappear after a certain age.

kingdom (for example, in baby birds learning their parents' songs), and the idea was that after a particular point in an organism's development, it became impossible to learn certain things. It was suggested that human children might experience something similar: If they didn't get the necessary second language input while they were young, the window would close, so to speak, and it would no longer be possible to learn the intricate system of a language fully from simple exposure later on.

Many studies have tried to address this question over the years, and researchers are still spilling quite a bit of ink on the details. Yet recent research provides reason to be optimistic about learning another language throughout childhood. For one thing, most researchers now talk about a series of "sensitive" periods, for different abilities such as accents, collocations of words, morphology and syntax, with phase-out periods, some at age six, and some in the mid-teens. They acknowledge that the brain changes throughout childhood, but they note that any decline in language learning ability that does occur is gradual, and it's not that the ability *disappears*; it just *changes*. In other words, there is no definite cut-off point (and there is hope for all!). Particularly in cases where there is strong motivation and positive attitudes toward the second language, older children and adults can make great strides, often learning more speedily than young children.

What Children Do with Language (and When)

So children tend to differ from adults in how they learn a *second* language. What's more, children also differ from each other in how and when they start using their *first* language. Researchers often talk about linguistic milestones, basically what we should expect from children as they begin to talk. The most basic milestones are babbling (before age one), reaching the one-word stage (at age one), and the two-word stage (at age one and a half or two). Two key points: First, children understand a lot more than they can produce. For example, a typical one-year-old child can understand roughly seventy words but can only produce about six! Second, there is *a lot* of variation across children. Some children hit their first word milestones earlier (say, nine months) and some much later (say, eighteen months).

SPOTLIGHT ON RESEARCH:

What Should We Expect from Our Children When They Begin to Talk?

Hundreds of research studies over the last decade give us a good sense of when we can expect what linguistically from our children. A child's linguistic development begins with the child experimenting and making sounds with no particular meanings attached. Although parents might be inclined to believe otherwise, very early crying is not, in fact, an *intentional* attempt to communicate! Later, when infants begin to coo, from approximately the second to the fifth month of life, parents often interpret this as a sign of the child's happiness (which it may well be!). After cooing, babies babble, making vowel sounds (aaaa-iiiiiii-ooooo), then repeated consonant-vowel sounds like "ma ma" or "ga ga," then more varied sequences like "ba ka du gu." Proud parents often interpret this early babbling as proof that their babies are already talking, or maybe have their own special language. While infants do put their vocal play to social uses (for instance, imitating the silly sounds that a caregiver might use with them), it is more likely that they are just playing with sounds and are not actually talking about specific objects or people.

Around an infant's first birthday is when he or she begins to produce sounds associated with specific meanings. For English-speaking children, these words are often nouns linked to everyday experiences (for example, *doggie, dada, milk*). At about age one and a half or two, children enter what is called the two-word stage and begin putting words together, such as *baby cry* or *eat now*.

Around age two, children go through what's known as a "vocabulary spurt," where they add about two hundred words a month to their productive vocabularies. This can be an exciting time as it may seem that all of a sudden your child has a word for everything! At the age of about two and a half, many children begin making phrases of three or more words, like *I need something* or *no want carrots* (both popular sentences in our houses). From this point onward, young children continue to add to their understanding of the grammatical rules of their language and to their vocabularies. By about the age of five, most children have mastered most of the grammar they hear regularly around them. Of course, there may still be some glitches (for example, mixing up vocabulary words by calling lizards *dinosaurs* or vice versa, or interpreting language literally—for example, understanding a "run-off" to be a running race), but these are part of the normal learning process.

Let's turn next to what parents need to think about in deciding when to introduce that second language, always keeping in mind the very important point that it's never too early or too late—there are pros and cons for all ages.

RAISING A CHILD BILINGUALLY FROM BIRTH: BABIES AND YOUNG TODDLERS

When children are exposed to two languages from birth, they'll view both of them as natural means of communication. However, if you didn't grow up bilingually yourself, you might be a bit nervous about introducing an additional language to your new baby. You might have questions like, If my baby is exposed to two languages from birth, is there any chance she'll get confused? Or if I expose my baby to two languages will she be a late talker? Leila and Ilya worried about these questions when thinking about their new baby. Leila was raised in the United States and her native language is English. She is married to a Ukrainian man, Ilya, who was raised in the Ukraine and speaks Ukrainian as his native language. He has lived in the United States for about twelve years and speaks English fluently, but with a strong accent. The couple now lives in Southern California. When we first heard about them, Leila and Ilya were expecting their first child in a few months and were talking about whether or not to raise the child bilingually from the outset. They had thought about using English only with their baby because a teacher friend had told them that very young children are too young to learn a second language. Her advice was that they should wait until the child was older and his English was firmly established. But at the same time, Leila and Ilya wanted the child to be able to communicate with Ilya's relatives as soon as possible (especially those relatives who speak only Ukrainian) and they didn't want to deny the child his Ukrainian heritage. For his part, Ilya would prefer to use his native tongue with the child but doesn't want Leila to feel left out (her Ukrainian skills are minimal). Mostly, he just wants to do what will be best for their baby.

Leila and Ilya knew that they wanted their child to speak both English and Ukrainian, which is a good start (the "which language" question was sorted out), but they worried about the best age to start. Let's take a closer look at Leila and Ilya's concerns. Some of their worries are probably familiar: What if our child, because of having to learn two languages, doesn't start speaking until later than other children? What if our child ends up not knowing either language perfectly? What if our child makes mistakes in one language and it seems like the mistakes are coming from the other language?

From our experiences in talking with new parents, we've found that concerns about language delay when introducing a second language to their newborn babies top the list—and in some cases are enough to be a deal breaker. And this isn't just a widespread concern among parents; we, and many other parents, have also heard it from doctors and teachers. That makes it even more important for us to emphasize the following: There is *no evidence* that learning more than one language from birth leads to delay. Children who learn two languages from infancy begin to babble, to say their first words, and to combine those words into two- and three-word mini-sentences at the same time as children learning only one language. There is a wide range of completely normal variation among children in how fast they learn language and *no credible research* out there that suggests speaking to a baby in more than one language is, in any way, detrimental. In other words, it is never too early to start with two languages!

So, what did Leila and Ilya decide to do? After the birth of their son, Nicholas, in 2001, they spoke both languages to him right from the start. Leila used English and Ilya spoke mostly Ukrainian. They continued to use mostly English between themselves, but they also found more and more Ukrainian making its way into their conversations. Nicholas made all sorts of learning mistakes in both languages when he was two and three years old and he often mixed languages at home. However, Leila and Ilya learned this was a normal part of the learning process, and when the family visited Ukraine, they found that Nicholas could speak (quite at length) to his relatives in Ukrainian and he could switch back to speaking English with his mother with ease. He entered kindergarten in the United States when he turned five

> **• FAST FACT •**
>
> It's never too early for your baby to start learning another language and learning a second language from birth does not lead to confusion. Bilingual children hit the same linguistic milestones at about the same time as monolingual children.

and is having no problems with either English or Ukrainian. Leila and Ilya are happy with their decision, and it allowed Nicholas to learn the languages of both his parents.

So, what's the best way to introduce another language to your baby or your young toddler? In the box below, we provide a range of tips for providing second language input.

 QUICK TIPS: Creating Good Language Learning Environments for Very Young Children (Ages 0–2)

• Direct lots of rich, meaningful speech toward your child from birth.
• Encourage friends, relatives, babysitters, siblings, and other visitors to speak and play with your child in the second language.
• Engage in interactions that pique your child's interests—for example, by using attention-grabbing toys, picture cards, or other props while you use the language.
• Build up positive associations by singing and dancing to silly songs, listening to music, and playing games in the language.
• Read stories to your child. Interact both with the book and with your child (for example, by acting out the stories and using funny faces and voices). Keep this light, fun, and brief.
• When you're looking for child care, find someone who speaks the second language (more on this in chapter 6).
• Be enthusiastic and have fun with the language!

INTRODUCING A SECOND LANGUAGE TO A YOUNG CHILD: PRESCHOOLERS

One of the wonderful things about preschoolers is that they make lots of mistakes as they learn how to do things, and *most* of them happily accept these mistakes as part of life. At this age, they are used to—slowly and with some trial and error—learning new routines, figuring out how their toys work, how to get the breakfast cereal they want, what a cubbyhole is for, how to put on their shoes, and so on. Young children are accustomed to not understanding everything. It's much the same with language, first or second. Preschool-age children usually don't worry about not being able to understand everything they're hearing. They are used to making mistakes when they are trying to ask for things and taking risks with a language, trying things out to see if they work. Think of all the baby books parents have kept that are filled with adorable phrases like "I holded the baby wabbits."

Still, if you are the parent of a toddler you might well wonder: *If our toddler is introduced to another language during the preschool years, will it be confusing?* In a nutshell: No. If children receive plenty of rich and varied opportunities to hear and produce the second language, introducing it in the preschool years won't be problematic, and the details will work themselves out. With toddlers, it's neither too early nor too late to start!

Parents like Daniel and Larissa figured this out on their own. Daniel and Larissa are native speakers of English, born and raised in Toledo, Ohio. Daniel is an engineer and was recently transferred to Saudi Arabia. He and his wife have a three-year-old daughter, Mary. Before they left the United States they wondered: *Should we try to teach Mary Arabic while abroad?* Daniel and Larissa didn't plan to live in Saudi Arabia for more than five years, but at the same time, they wanted Mary to be able to communicate with children who speak Arabic and to have the benefit of learning another language. Should they invest in Arabic classes for her? Or should they just focus on getting settled and making friends with the British and American ex-pats in their community?

Daniel and Larissa heard from their neighbors and co-workers that "the younger the better" is the rule for learning another language. One of Daniel's colleagues told him you had to start learning a foreign language before the age of six to be able to master it. He gave himself as an example, saying that he had started learning Arabic when he was thirty, and that now, after ten years, he still struggled.

Was Daniel's colleague right? Well, he was partially right. As we've said, you don't have to learn a foreign language before a certain age to master it, but if you want to have a nativelike accent, then hearing the sounds of that language early on certainly seems to help. Researchers tend to agree that younger is almost always better for accents. However, there's sure to be more to why Daniel's colleague has not been successful in his learning than the age when he started. We talk about the effect of things like aptitude later on in this book. However, for now, let's ask what researchers say. Is younger always better?

SPOTLIGHT ON RESEARCH:

The Younger the Better?

In 1978, researchers Snow and Hoefnagel-Höhle studied native English speakers in Holland who were learning Dutch. They followed learners of different ages (preschoolers to adolescents to adults) for about a year, testing them every four or five months on a wide range of skills (for instance, pronunciation, the ability to distinguish different kinds of language sounds, vocabulary, and grammar tests). Interestingly, the researchers found that the adolescent learners did the best on most tasks as far as *rate* of learning was concerned. Adolescents learned more Dutch—and learned it more quickly—than either adults or young children!

In short, researchers have concluded that younger is better for accents, but this is not necessarily always the case for some of the other things that make up knowing a second language, such as vocabulary and grammar. Why is younger better for children's accents? It's probably because younger children still have the ability to perceive and discriminate between many new sounds (such as those in the second language). In contrast, older learners are more likely to have problems hearing new second language sounds, instead trying to make sense of them by using the sounds they already have (and have been using for years) in their first language. For example, in Hungarian there is a sound that, to a native speaker of English, sounds somewhat like a "guh". It's not really the same as "guh" (it's produced farther forward in the mouth), but native speakers of English learning Hungarian tend to treat this unfamiliar sound as a "guh" sound. In other words, they categorize (and pronounce) the new sound based on a sound they already know.

To overcome this tendency to use their existing sound systems, older learners have to reorganize their entire sound systems. This is no small task, and focused training is generally needed to help older learners to pay attention to sound contrasts that don't occur in their first language. Babies and younger learners, however, don't usually need this type of training. Their brains and sound systems are still flexible enough to be able to hear and process sounds like native speakers do. Of course, the amount and intensity of a learner's interactions in the second language also play an important role.

• FAST FACT •

Younger children often still have the ability to perceive and discriminate between many new sounds in the second language; older learners tend to make sense of these new sounds by using their first language.

So, what happened with Daniel, Larissa, and Mary? They moved to Saudi Arabia and have been there for two years. While they are continuing to speak only English at home, the entire family is studying Arabic daily with a tutor, and even using it a bit in everyday life. They have also made an effort to get to know people and to become involved in the community outside their home. They found that Mary began picking up sounds and words fairly quickly and is having a fun time playing around with pronunciation. She is not forgetting English and enjoys playing with girls her age and a bit older in the women's language exchange circle that her mother attends. Larissa is struggling a bit with the pronunciation (Arabic has some sounds that English doesn't have), but she's making progress. Daniel is motivated to learn the language and enjoys being able to communicate and even occasionally understand the jokes his colleagues make. Daniel and Larissa are glad that their child is learning Arabic, and it's helped to make their stay there more meaningful and interesting in many ways.

Another excellent thing for language learning in relation to preschoolers is that this is the age when many children begin to love books. Today's parents understand that reading to their children is crucial because it helps them with everything from vocabulary development to literacy skills and bodes well for academic success. Reading to your child in the second language provides the same kinds of benefits. Finding books in the second language that are appealing, interesting, and useful is an increasingly easy and enjoyable task with all of the online bookstores that now exist. (See References and Resources for more information.)

 QUICK TIPS: Creating Good Language Learning Environments for Preschoolers

- Have fun integrating the language as a part of your daily routines. For example, sing morning songs in the language, play alphabet and counting games, guessing games, and have a word of the day.
- Read stories to your child in the language. Keep these light, fun, and brief. Encourage your child to interact with the book and you

(for example, together, act out the stories, use funny faces, give voice to the characters).

- Find other children who speak the target language for your child to play with. Make these language dates fun by providing props (treats and toys, musical instruments, scavenger hunts). Children learn a lot from each other. Even finding children a bit older than your child will provide positive "big boy" role models.
- Look for games in the target language, including things like board games and flash cards that encourage interactions.
- Find funny cartoons and characters that use the target language.
- Use crafts as an opportunity to speak and interact in the target language. Consider making cultural learning opportunities out of the craft time as well.
- Play songs in the second language in the car or use headphones on public transport.
- Be enthusiastic and positive about learning the language.
- Don't be overly focused on perfection or correction, instead focus on what your child has achieved.

LANGUAGE LEARNING FOR SCHOOL-AGE CHILDREN

Whereas young children often appear to soak up languages effortlessly (including things like pesky verb conjugations), older learners seem to have the ability to learn faster, although these speed differences usually wash out in the end. Aptitude also plays a greater role for older children than it does for young children (although this might be because we can measure it better as children get older), probably explaining at least part of the reason why two older children can start learning the same language at the same age with quite different results five years down the line. For school-age children, it is important to capitalize on the strengths that they have. For example, unlike little ones, older children have the ability to understand or even occasionally talk about some grammatical rules. Plenty of people who started learning a language after early childhood consider themselves quite successful and are fully functional in their second language.

Children at school are rapidly acquiring lots of academic skills that can help them learn a second language. For instance, learning to read not only promotes the development of awareness of second language grammar, but it also provides an additional way of understanding and remembering a second language. School-age children can make use of their amazing sponge brains *along with* some of the adult-like mental tools they're starting to pick up for a variety of learning tasks. We often hear questions from parents about language skills and reading development. For instance, *if a second language is introduced in early elementary school, will it interfere with my child's developing literacy skills?* As a matter of fact, bilingualism, language awareness, and literacy are positively related. So, is it too early or too late to introduce another language to children in elementary school? Well, no!

By the time children are in middle school and then high school, they have developed some pretty sophisticated cognitive skills and abilities. They begin to learn things like algebra and computer programming, and some of these analytical skills can come in handy for the task of learning the grammar of a second language. At the same time, parents may wonder: *Middle school is such a stressful time as it is. Will s/he be self-conscious about making mistakes in another language?* And while some parents can't remember how to speak French from their high school French classes, they often do remember how hard they struggled to memorize all those verb conjugations. A common question is whether high school kids nowadays will likely be subject to the same fate. Admittedly, we can't allay all of parents' fears here. We wish language instruction or exposure could start before high school for all children. And if it can't, we hope there are plenty of opportunities, through summer camps and extracurriculars to supplement language study at school. If parents encourage good attitudes toward another language, this can go a long way toward making language learning in older childhood an extremely rewarding experience. So, while older children don't learn language in exactly the same effortless way that younger children do, this doesn't mean it's too late for older children; it just means that different approaches and strategies tend to be more helpful. *Is it too late to start to learn another language in middle or high school?* You guessed it—no!

What are these helpful strategies? Most older children benefit from some explicit attention to grammar and vocabulary. However, this doesn't (we repeat: *does not*) mean that the best way for older children to pick up a language is by endlessly studying boring lists of grammar rules. Meaningful communicative interactions are important for learners of any age. It's just that as children get older, we need to make the best use of their own learning styles and resources for optimal outcomes. That sometimes involves supplementing the types of input, exposure, and interaction opportunities they receive in different ways (for example, by making sure older learners get some feedback on their levels of accuracy by setting up language exchanges each week). Correction, which can be useless or discouraging for younger children, can be helpful for older children, especially if done carefully. (As many of today's teachers know, correcting every error can be counterproductive. Correction and praise need to be balanced.) Older children, with their greater ability to be analytical, usually find correction more helpful.

For younger children, little linguistic details tend eventually to fall in line with how native speakers would use them. For older children, on the other hand, even if they truly want to sound nativelike, they still sometimes leave off bits of language. This is particularly true for linguistic elements that don't make much of a difference for meaning, like the third-person singular -*s* in English (as in *she walks* versus *they walk*). The -*s* at the end of walk doesn't contain any particularly meaningful information, as we already know that the subject (she) is singular and third person. These

> **• FAST FACT •**
>
> Older children often need a bit more support and encouragement, and can benefit more from explicit instruction.

sorts of errors are not the fate of all older learners, but it is important to be aware that this is more prevalent in older children than in young ones.

Ideally, you should try to provide your children with opportunities to speak and hear lots and lots of the second language, no matter what age they are. But, if you decide to introduce the language when your child is older, providing additional opportunities to pay attention to the grammar of the language is important, too. So, while playgroups and plenty of language input might be fine for toddlers, your school-age child, especially if they are in middle or high school, is likely to benefit from formal classes as well.

SPOTLIGHT ON RESEARCH:

Can My School-Age Child Ever Sound Native?

Most researchers agree it's *possible* for some to learn to speak a language without a noticeable accent when they began learning it after early childhood, but it's probably the exception rather than the rule. For instance, in one study from 1999, Flege, Yeni-Komshian, and Liu tested the English of 240 Korean immigrants to the United States in order to assess how nativelike their accents were. All of the subjects had lived in the United States for at least eight years and had an average of ten years of education in the country. Their ages of arrival to the States ranged from one to twenty-three years old.

The researcher found that *overall* the later the Koreans had arrived, the stronger their foreign or Korean accents were in English. Well, that doesn't sound too promising, you might be thinking. It's not all bad though. For instance, *a few* of these Koreans did sound like native speakers of English. So it is possible for some older learners to develop great accents no matter what age they start!

So, why do some late beginners manage to sound so good? Research suggests that these star learners (who begin late but can pass as native speakers) have achieved their level of ability through a combination of (1) large amounts of language input; (2) high motivation; (3) natural aptitude; and (4) intensive training. We discuss things like motivation and aptitude in our chapter on individual differences, and intensive training in our chapter on language programs and teachers. The relevant point for this chapter is that if you decide to introduce a second language to your school-age children, it's not impossible for them to develop a native accent, *but* if that's the goal, they'll need a lot of high-quality language input, and certainly more than the once- or twice-a-week class offered by some public schools.

"How Important Is an Accent, Anyway?"

With these sorts of research findings in mind, another thing to ask yourself is how important it is for your child to have a nativelike accent. Speaking a certain language is connected to taking on a certain identity. And all children will develop their own feelings about how much they want to adopt the identity that goes along with speaking a particular language and sounding nativelike. An accent can be seen as a marker of a person's background, and some people find that they like and appreciate that! In fact, since learning another language can be a healthy part of building a self-concept, parents can use the language learning process as a way of letting their older children explore and develop their own identities. That may mean that one child wants to fit in and assimilate to a particular group as a full member, while another wants to assert his unique identity as someone moving between two groups. It will be fascinating to watch their accents develop (and change quickly depending on where they are and whom they are with).

It's also worth keeping in mind here some basics of social and emotional development because these factors become more important at this age. Most psychologists would say that before the age of about eight, children usually haven't developed a strong sense of self-evaluation. In other words, they don't critically evaluate or second-guess themselves. As they get better at taking other people's points of view into account, they also start to worry about other people's perceptions. By the time they're eleven or twelve, self-esteem sometimes decreases as children start making more critical evaluations of themselves. These changes in self-concept can also be linked to motivations for learning languages. For instance, younger learners might seem fearless about being immersed in the second language (through, say, an elementary school immersion program); they may thrive on the social interaction and jump right in (even if that means making lots of mistakes). In contrast, a twelve-year-old learner might be much shyer and even intimidated by such an experience and would benefit from some explicit instruction and some specific strategies right from the start to help feel a bit more confident.

> **· FAST FACT ·**
>
> *Sounding exactly like a native speaker of the target language is less important than being able to communicate in the language and enjoying using the language.*

QUICK TIPS: Creating an Optimal Second Language Learning Environment for Your School-Age Child

- Have fun with the language. For example, encourage puns, riddles, jokes, and cartoons in the second language.

- Read with your child in the language, or better yet, have your child read to you. For older children, encourage them to choose favorite books and then read and discuss them together.
- Find children who speak the target language for your child to play and interact with. Make these language dates fun! For example, you can amass a set of board and card games and bring them along.
- Get a newspaper in the foreign language, cut out headlines and discuss them.
- Have a regular activity in the target language (such as cooking a certain dish together).
- Find some music or an artist you both like, buy the CDs or download the files from a music store and listen to them regularly, discussing what's there. There are a number of bilingual artists (like Shakira) who appeal to school-age children.
- Try to keep them interested and engaged by varying things and mixing it up a bit. Every week, find a new activity, even if it's a movie, that involves the second language.
- Encourage your children to find second language pen pals. Lots of Web sites can help you find a good match.
- If you have family in the country where the second language is spoken, try to either plan a trip there together or learn all about that country by surfing the Web together. Keep a folder, add to it regularly. There's nothing like an actual trip to really motivate a language learner, and experiencing the language in a real, living environment is priceless!
- Be enthusiastic and positive about learning the language. Minimize stress.
- Don't put too much emphasis on having a perfect outcome.

WRAP UP

Many parents worry about when they should introduce a second or foreign language to their child. The take-home message of this chapter, however, is that it's never too early or too late to start learning a new language. The best time to start is right now!

Perhaps one of the most important things for parents to do is to tailor the learning environment and strategies to the ages of their children. Young children do well when they are given lots of language input to play around with, so you don't need to worry about grammatical rules for three-year-olds. It's much more important to engage them with something fun and interesting (and consistent) in the second language. Older children, on the other hand, do best if their language interaction opportunities (which are still highly important) are supplemented with some explicit approaches for focusing on particular features of language. They can make use of their superior capacities for abstract thought and analysis, but this should always be in *meaningful* contexts. They may also need to be motivated in different ways. So, practically speaking, this means that we should all be focusing more on making language learning a positive, fun, interactive, and engaging process and worrying a bit less about what we've heard kids can and can't do at certain ages.

EXERCISE:

*Brainstorming Ideas for Creating
a Positive Learning Environment for
Your Child Right Now*

We discussed in this chapter how important it is to help your child develop a positive attitude and high motivation for learning the target language no matter what his or her age. Take a minute to brainstorm your ideas for ways to do this. Sample answers are at the back of the book. (See Resources.)

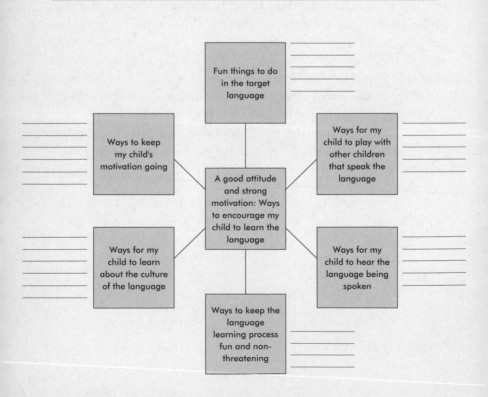

CHAPTER 5

How Do Individual Differences Like Birth Order, Gender, Personality, Aptitude, and Learning Style Affect Your Child's Language Learning?

In this chapter, we talk about the role of four factors that might impact your child's second language learning: The big four are birth order, gender, personality, and language aptitude. After explaining how these individual differences can impact language learning, we provide practical tips to help you leverage your own children's strengths for better learning. By the close of this chapter you'll have a better idea about what makes your child unique as a second language learner and how you can help maximize these individual characteristics to benefit her or his language learning.

DOES BIRTH ORDER IMPACT SECOND LANGUAGE LEARNING?

What if a younger child doesn't seem to be learning as quickly as an older child did, or vice versa? Is this natural or something to worry about? Questions about birth order are shared by many monolingual as well as bilingual families. For instance, Jakob and Magdalena have two sons: Zack and Hans. As native speakers of German who have lived in the United States for many years, Magdalena and Jakob are raising their sons bilingually (in German and English) because they believe this will give their children a head start in life. However, they have noticed some patterns that surprise them. Zack, their elder son (now three and a half), began speaking at twelve months and had a large vocabulary in both German and English by the age of two. Hans (who just turned two) seems to be a late bloomer, and his vocabulary in both languages is still relatively small. Jakob and Magdalena worry: Why is there such a difference between their two sons in terms of their language development? Will Hans ever catch up? Is there anything they can do?

Researchers have found that for *first* language learning, first-born children have an edge. Later-born siblings tend to have slightly smaller vocabularies than their older brothers and sisters had at the same age, and they add new words to their vocabularies at slower rates. Why do first-born children outpace their siblings? First of all, they get more time. Simply put, the more children there are in a family, the less one-on-one time each child gets with parents. All eldest children have a period of time when they are the center of attention (and for only children this golden period does not end). Later-born children have to share the spotlight with their older sibling(s) right from the start. Research indicates there is a small (but consistent) relationship between the amount of time children have with their parents and the size of their vocabularies. These small time and learning differences between older and younger siblings in a single family should not be a cause for worry, though— they get ironed out quickly.

Another reason is what *sorts* of language these first-born children hear. First-born and only children have usually played lots of name games ("Where's your nose?" "Is this a doggy or a piggy?") with their parents, so they quickly develop a lot of nouns in their vocabularies for labeling things. In contrast, younger children hear different kinds of things from their parents and siblings: "Hey, you guys!"; "Lemme see!"; and "Stop it!" This is common sense if you think about it. Both parents *and* older siblings have a big impact on the language environments of later-born children. First-born children get more individual time with their parents; later-born children often hear conversations involving more than two people. As a result, later-born children may develop some conversational skills earlier than their older siblings did, like how to interrupt and get everyone's attention and how to understand who people are talking about when they use pronouns like "he" and "she."

This advantage for older children has been documented for *first* language learning, but what about for *second* languages? In a 2002 report, researcher Sarah Shin investigated Korean-speaking families living in the U.S. with more than one child and how they used Korean and English in their homes. For similar reasons—mainly that parents have less one-on-one time with later-born children (in this case, one-on-one time in Korean)—the first-born children were more bilingual in Korean and English than their younger siblings.

 SPOTLIGHT ON RESEARCH:

Second Language Use in Families with More Than One Child

Shin investigated the language experiences of second-generation immigrant Korean American school-age children (four to eighteen years of age) by surveying 204 families. Shin reported big differences across siblings according to birth order. She found 79 percent of first-born children, 66 percent of second-born children, but only 43 percent of third-born children spoke Korean with their parents. First-born immigrant children received more direct speech input from their parents than other children. She recommends older children be encouraged to help their less proficient siblings as they speak the second language, and that parents be made aware that they might speak *less overall*, and *less in the second language*, to later-born children.

WHAT HAPPENS WHEN OLDER CHILDREN GO TO SCHOOL?

As soon as children begin school, they are more inclined to speak the language of *that* new and fun environment. Many parents in the United States who are trying to promote a non-English language at home describe this as a key transition period. For instance, Nasim, a mother of two children—Nika and Andisheh—describes her efforts to maintain Persian as a heritage language. While she was successful in promoting the language for her first child, it became more difficult when the second daughter came along. As she explains, "With my first daughter, Nika, I did not put her into preschool until she was three and a half, so those three and a half years it was almost pure Persian, because there was no one else to sort of pollute the environment. So she'd hear only Persian or Azeri. Whereas with [my younger daughter] Andisheh, Nika's friends would come, they are speaking English, and so Andisheh is responding in English. Everything just was more lax with Andisheh than it was with Nika."

When an older sibling goes to school, it becomes much harder to limit the amount of English used at home. When schoolwork and activities come into the mix, it is sometimes hard to find the time for the same activities and events that were associated with the second language in the home. Because older children often return home from school speaking more English, they can change the language balance for their younger siblings.

Recognizing Younger Children Might Need a Bit of Extra Help

It can be pretty easy and tempting for younger children to follow the lead of older children. If the goal is for children to acquire a minority language (that is, a language that's not commonly used outside the home), you might have to work a bit harder at ensuring rich input in that language for your younger children. You can use some of the time that the older child is away at school for language-rich activities such as a language-based parent-and-child play-group or to squeeze in some extra one-on-one book reading time (see chapter 6). You might also want to consider setting up some family rules about language use and talking with your older child about her important "big kid" role in teaching a younger sibling the second language. Depending on the siblings' relationships, you might also promote the home language as a secret code they can use only among themselves that won't be understood by the other kids. You can also foster the second language as something desirable for younger children to do, just like their big sister can tie her shoelaces. As parents, we all have ideas about how to motivate our children for things like toilet training and bedtimes; leveraging these tricks for language learning takes a little extra effort and ingenuity.

 QUICK TIPS:

- Whenever possible, try to spend at least a little one-on-one time with each child. This can be something as simple as reading a quick story, playing a game, or taking a trip to the grocery store together. Make this a special time to practice language skills. When older children go to school, make the alone time especially language rich for younger children.

- Try to use specific and descriptive words with your children instead

of general all-purpose words like *thing* and *do*. Elaborate (and repeat).

- Don't worry about holding younger and older children to *exactly*
the same standards. If the younger child is slower at developing,
research shows that they usually catch up.
- If older siblings take on the role of translators for their younger
brothers or sisters, encourage them to let the younger children
speak for themselves. But also encourage them in their role as language tutor.
- Trust your instincts. You know your children. If one does seem to be
progressing *significantly* more slowly than you think is right, ask a
language development professional (like a speech-language pa-
thologist at a local elementary school) about it.

ARE GIRLS OR BOYS BETTER AT
SECOND LANGUAGE LEARNING?

Most parents notice big differences based on gender right away. For
instance, who knew—certainly not us—that by their first birthday,
many boys would gravitate toward things like toy trains or trucks,
and girls toward princess dolls or cuddly toys, like high-powered
magnets! There are also interesting language and learning differences between girls and boys that parents need to understand. Researchers have found that girls generally begin talking a bit earlier
than boys, and that girls' vocabularies grow more quickly than
boys' do. Studies focusing on more subtle and specific linguistic
skills, for example the complexity of children's speech (sometimes
measured by counting and then averaging the number of words in
children's sentences: for example, "Mine!" [one word] or, "I need
my blue train right now." [seven words]), have found some differences there as well, again with advantages for girls. Researchers
have also found a slight advantage for girls in vocabulary production and comprehension.

These differences between boys and girls might be due to the fact that girls grow up more quickly in general. For instance, research has found girls' brains mature more quickly and in slightly different ways. Infant girls, for example, typically have longer attention spans than infant boys. The ability to pay attention to things in the environment (for example, listening to people talking, and looking at what they are looking at) is important for language development. (After all, if you are not listening, it's hard to learn!) Researchers have also found that newborn girls pay more attention to their mothers' faces than boys do. So, girls may be more likely to watch their mothers as they talk. These biological differences seem to appear right from the start, and other than being aware of them, as parents, we probably can't do much to change them.

 SPOTLIGHT ON RESEARCH:

Gender Differences in Vocabulary Development in the First Language

In 2002, researcher Daniel Bauer and his colleagues investigated whether there are differences between boys and girls with respect to their production and comprehension of vocabulary words (that is, in the words they can say and the words they can understand). Twenty-six children, all being raised in English-speaking homes, were assessed monthly (from about eight months up through fourteen months of age). Parents completed vocabulary checklists on which they marked off the words their children understood as well as the words they both understood and produced. Overall, they found a slight advantage for girls in *both* vocabulary production and comprehension. Bauer and his colleagues speculated that the differences might be due to the fact that girls mature more quickly than boys do in general.

It is important to remember that these language and learning differences between boys and girls in the early years are overall quite small and are usually not hugely significant as children get older. There may well be larger differences between a particular pair of girls, due for instance to things like their personality, birth order, or aptitude for learning, than there are between a particular girl-boy pair. And when you look at the big picture, girls and boys are a lot more similar in their overall language development than they are different. So, as parents we need to recognize that each child is an individual and will proceed at his or her own pace.

Creating Conditions for Successful Second Language Learning for Girls and Boys

There are many ways that parents can leverage these gender differences to provide optimal learning conditions for both boys and girls. Researchers report lots of differences in how boys and girls are treated from a young age. Some investigations, for example, have found that girl babies hear *more* parental talk in the early years than boys, and that mothers tend to produce *more complex* sentences and ask more open-ended questions (for example, "Why do you think so?"; "What should we do today?") of their daughters compared to their sons. If parents believe that their girls are social and cooperative, they may put extra effort into finding playmates for them and encouraging them to interact and play together. This is helpful for their language development.

Other research reports that parents' conversations with their boy toddlers involve play situations, where the focus is on concrete events happening at the moment, while their conversations with little girls tend to focus on more abstract concepts not present in the immediate context. All of this suggests that some of the language differences we see between girls and boys might be due to the fact that (often despite our best intentions) we treat boys and girls differently from very early on. Once we realize how these things can impact the environment for language learning, we can work to try to balance our talk to girls and boys—both in first and additional languages. This means that you should take your child's gender together with her particular interests into account in devising optimal second language learning strategies. For instance, in promoting a second language, we want to make it enjoyable. We want to interact verbally with our children in ways that engage them. Girls and boys often have different interests. For example, researchers have found that girls enjoyed talking more about personal topics (such as staying over at their friends' houses). In contrast, boys liked to talk about breaking rules or daring activities. Obviously, it is important to find topics that interest your child as an individual (whether it's your daughter who loves motorcycles or your son who loves to cook!) than it is to follow generalizations or rules about what boys and girls talk about. One way of doing this is to focus on second language topics and areas that your child is enthusiastic about in his or her first language, recognizing that it might be easier to engage your son with the here and now of his dump trucks, and your daughter with the relationships between her stuffed animals at her tea parties. Or vice versa. Finally, you should keep in mind that rich, complex interaction is helpful for both boys and girls, as are playmates and social situations.

PERSONALITY AND SECOND LANGUAGE LEARNING

From the moment babies make their appearances, we parents begin to look for clues to their personalities. We often interpret any sigh, smirk, or screech in those early weeks as a sign of how easygoing, funny, or stubborn a child is. Personality differences tend to become more pronounced as children grow. By age three, your child may be talking a mile a minute and always seeking something new. Or you might have a child who chooses his words very carefully and prefers quiet, predictable activities at home. Some children keep an even keel—others are a bit more sensitive. Do any of these differences affect second language development?

Fernando and Bernice have similar questions regarding their three-year-old boy Malcolm, who they are raising bilingually in Spanish and English. Mal is a quiet boy who prefers to play by himself. He is easily distressed by changes in his schedule, new events, and new people. He would rather play with his train set than talk. Fernando and Bernice have noticed that Mal doesn't talk as much as other boys his age, and he seems to have a smaller vocabulary. Is all this due to his temperament, they wonder? Should they be worried?

Your child's temperament is made up of lots of different components. These include how your child relates to others (some children are outgoing, others are more introverted), how your child responds to new or potentially distressing events like a first haircut (uncontrollable crying? relaxed and easygoing? hyperstimulated?), and how long your child is able to pay attention to particular objects and people. To a large extent, these are relatively fixed, in-born traits that influence how the child approaches the world, including how the child goes about *learning* about the world. There's not much we can do as parents to change our child's basic temperament.

Partly for this reason, different temperaments shouldn't really be thought of as good or bad on their own. It's only in relation to particular situations that particular temperaments might be more or less favorable. For instance, children who have trouble sitting still might be exhausting for parents, but that high activity level might contribute to making them successful in sports at school and high-energy careers later in life. Another example: Persistent children who continue doing what they're doing in the face of adversity might be criticized as stubborn in a particular context, or praised as patient if working on a tough problem.

As you might guess, a child's temperament can play an important role in second language development. For example, if your child is outgoing, she might seek out others more often for conversation and play; this in turn provides her with more exposure to the language and more opportunities to use the language. On the other hand, if your child is shy, she might be less likely to take risks and might have long silent or quiet periods when learning a second language.

Researchers sometimes distinguish between social and cognitive strategies used by second language learners. For example, social strategies include joining a group and acting like you know what's going on (even if you don't); giving the impression you speak the language with a few well-chosen words; and counting on your friends for help. Comparable cognitive strategies are: assume what people are saying is relevant to the topic at hand (i.e., guess what people are talking about); look for recurring parts in expressions that you know and use them (for instance, "How do you [find the train station]?"); and make the most of what you've got (say what you can and don't worry about details).

Lily Wong-Fillmore showed how these strategies worked together to help learners appear to be competent and also to attain competence in a second language. Children who are introverted may have a hard time jumping into a new playgroup and pretending that they understand everything that's going on, picking up the necessary language as they go. However, parents are able to help shy children by letting them observe at a comfortable distance, then by talking about specific strategies and phrases they can use to join in (for instance, "Look! Some cars! Let's have a race!").

Sensitive Children and Second Language Learning

Children who are sensitive or more emotional may sometimes have fewer cognitive resources available for processing language if they're using up a lot of mental energy being anxious. Learning language involves paying attention to what others are doing: What words do they use? When do they use them? If your child is easily upset, she may not have the mental space to pay attention to these kinds of things. (Think about when you are upset, for example, about having a fender bender, and right afterward you have to exchange all your insurance details with the other driver while also worrying about your children in your backseat. It's hard!). So it's not a surprise that children who are less easily rattled tend to find themselves in situations where they can pay better attention to what people are saying. As a result, they may experience quicker language development.

Sensitive children can be supported by providing familiar and supportive routines and pressure-free situations to experience and enjoy the second language. Predictability is key here. Sensitive children may shy away from bilingual playgroups but love to cuddle on the sofa with a book and a parent, or listen and sing songs with the CD in the car. Introduce new situations and people (like a bilingual playdate) little by little and with lots of discussion beforehand. Be aware that your shy and thoughtful child could be listening from the sidelines and remembering much more efficiently than the confident child who jumps right in and forgets a lot of what they are hearing.

Parents, Temperaments, and Second Language Learning

So research tells us that your child's temperament is part of the language learning equation. Another critical part is your reaction to your child's temperament. There is complex interaction between a child's temperament, the way *parents* respond to their children's temperaments, and language development. A very simple example: If parents regularly react to children's tantrums or even refusals to speak a second language with either time-outs or by refusing to engage with the child, this may deprive the children of the types of social interactions that promote language development. On the other hand, if parents respond by talking to their children in a soothing manner, the continued social interactions can provide the children with language input and a chance to talk—both of which may help support early language development. Similarly, children with responsive and supportive parents may feel more secure about exploring their environments—including talking to others.

We need to remember, of course, that sometimes you can't avoid a time-out, and also, temperament alone does not determine the course of language development. Language learning tendencies and opportunities interact with a wide variety of factors, including birth order, gender, other abilities, and aptitudes, as well as the parenting styles of the child's caregivers. Still, you'll want to consider your child's temperament in devising optimal language strategies. Answers to questions such as *How easily is your child frustrated? For how long can your child sit down and stay focused on one task? How does your child react to challenging situations? How willing is your child to take risks?*—are important in making optimal language choices for your child.

WHAT ROLE DOES MY CHILD'S APTITUDE PLAY?

Adults frequently talk about people who have a certain knack for language learning. Many of us remember being in language classes where some students seemed to breeze through each new exercise (like Kendall) and others who struggled every time (like Alison). But what exactly is language learning aptitude? And more importantly, does aptitude play a role in *child second* language learning?

While some researchers believe that aptitude doesn't play much of a role in second language learning by children, others disagree. For example, British researcher Peter Skehan looked at the relationship between first language development, aptitude, and second language achievement among British children learning French and German in secondary school. Skehan found a significant relationship between aptitude and second language learning. So, the jury is not quite decided concerning the role of aptitude in second language learning among children. However, it seems likely that the older the child is, the more likely aptitude is to play a part.

Aptitude and Your Child's Age and Second Language Learning

Aptitude is a fuzzy concept when we are considering young children who are still developing cognitively and acquiring lots of skills very quickly. Many researchers agree young children learn their first language through processing the language they experience in everyday interactions with other humans, relying explicitly and exclusively on social and cognitive skills. So, younger children probably don't need the same kind of explicit learning skills that adults do, for example, the ability to decode grammar. These analytical sorts of skills may become important if children start learning in their teens, or when they are trying to learn to read and write or express more abstract ideas in a second language. As we discussed in the "When" chapter, be ready to provide extra support and time if your child is starting a little later and hasn't used both languages consistently since birth.

Researchers have identified one particular aspect of aptitude that seems to play an important role in child language development. This is called phonological working memory, and describes the ability to remember (for short periods of time) what you've just heard. For example, if someone tells you a phone number and you have a good phonological working memory, you can remember that phone number (at least for a little while, and particularly if you repeat it to yourself!). There has been a lot of research on this topic in recent years, and it seems that children with higher levels of ability in this area (i.e., those who can remember what they've heard) experience quicker language development.

Children May Be Holistic Learners or Itemizers

Some children, early on, seem to focus on relatively small segments of language like syllables, whereas other children appear to acquire phraselike utterances as chunks with adultlike intonational patterns. Some children establish fairly regular and consistent sound systems, sometimes even resisting the imitation of new words that don't go along with what they're able to produce sound-wise, whereas other children are a bit more willing to take risks with new sounds, even if it means that their pronunciation ends up sounding a little bizarre. It's as though some children choose fluency, while others choose precision, and there may initially be an inverse relationship between the two. (For example, you can have one or the other—but not both!) This is not something for parents to worry about, though—just something to be aware of as they watch their children learn, and to encourage children who appear strongly oriented one way or the other to try something new.

Children May Be Auditory, Visual, Kinesthetic, or Tactile Learners

Children may display preferences for a particular learning style (or styles). Auditory learners prefer to hear the language input, so things like listening to songs and stories is very important for them. Visual learners like to see language input, so looking at words and pictures and making connections are how they enjoy learning. For kinesthetic learners, movement helps them, so, sports, field trips, and movement classes are all helpful for them. Tactile learners respond to touch, so finger painting and cooking might help them remember. What is important for parents is to understand the differences between these sorts of learning styles and preferences and, especially in families where there is more than one child to consider, to mix it up a bit, and use a number of different kinds of techniques as they select activities and materials. The diagram on the following page provides some more helpful hints.

Suggested Language Learning Activities Depending on How Your Child Learns Best

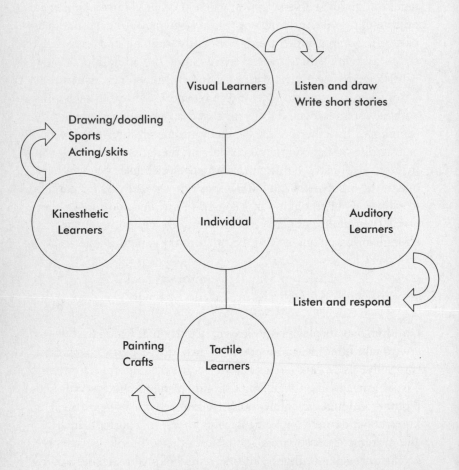

Multiple Intelligences

A related point: while two children will likely have different pre-ferred learning styles, they also likely have different types of intel-ligence. Psychologist Howard Gardner has identified eight different intelligences, and tests of these intelligences have been created for children as young as eight. Understanding a little about these can help in understanding your child's strengths, style, and preference for learning. The intelligences are described as being like talents and combinations of more than one are possible.

Gardner's eight intelligences are:

- Linguistic—the ability to use language to describe events, to build trust and rapport, to develop logical arguments and use rhetoric, or to be expressive and metaphoric.
- Logical-mathematical—the ability to use numbers to compute and describe, to be sensitive to the patterns, symmetry, logic, and aesthetics of mathematics, and to solve problems in design and modeling.
- Musical—the ability to understand and develop musical technique, to respond emotionally to music, and to create imaginative and expressive performances and compositions.
- Spatial—the ability to perceive and represent the visual-spatial world accurately, to interpret and graphically represent visual or spatial ideas.
- Bodily-kinesthetic—the ability to use the body and tools to take effective action, to appreciate the aesthetics of the body.
- Interpersonal—the ability to organize people and to communicate clearly what needs to be done, to discriminate and interpret among different kinds of interpersonal clues.
- Intrapersonal—the ability to assess one's own strengths, weaknesses, talents, and interests and use them to set goals.
- Naturalist—the ability to recognize and classify plants, minerals, and animals, and to recognize cultural artifacts like cars and sneakers.

It is important to realize that if your child displays a strong tendency toward one type of intelligence or another, it can be helpful to play to their strengths and provide language exposure that matches these intelligences. For example, songs in the second language for the musical learner; grammar activities for the linguistic learner; cooking, sports, and movement classes for the bodily-kinesthetic learner; field trips and arts for the spatial learner; language exchanges for the interpersonal learner; and so on.

WRAP UP

In an age of high-stakes testing and competitive parenting, there are intense pressures on parents (and children) to get into good schools, pass every test, and be "the best." However, in relation to second language learning, instead of comparing your child with others (and creating stress for you and your child in the process), we hope this chapter has helped you to value your child as he or she is: a unique individual—and a unique language learner!

See page 268 for an exercise that will help you keep track of and celebrate your child's achievements.

How?

CHAPTER 6

How Can You Best Promote Language Learning at Home?

anguage learning—like all learning—begins at home. And language learning—again, like all learning—works best when it's enjoyable for everyone involved, integrated into everyday routines and interactions, and meaningfully connected with real life. These points are the guiding principles of our approach and are well supported by a vast array of research from very different fields of study. In this chapter we integrate key research findings on language learning in different sorts of families with practical tips on child care, playgroups, and grandparents to help you create the optimal second language learning environment for your child in your home.

HOW CHILDREN LEARN TO USE ANY LANGUAGE

Children learn language through daily contacts, emotional bonds, and everyday interactions with their caregivers. Language learning happens, for instance, while children are playing everyday games like peekaboo and being talked to while having their clothing changed, sitting in their car seat, and being fed and bathed. Through thousands of these interactions over many months, young children gradually learn about the role of language in social life. And eventually they begin to recognize the language patterns that are produced within these interactions and become able to participate in them.

Here's the fascinating part: As children become more adept at communicating and participating in these day-to-day interactions, their caregivers naturally begin to use more complex language forms with them. The interactions between caregiver and child gradually become more complicated and sophisticated. In this way, the caregiver supports or scaffolds the child's emerging ability to speak.

And—there's no big surprise here—children's own language development is closely linked to their parents' language. For instance, researchers who have examined vocabulary growth have quantified the sort of language input children receive (for example, the number of words per hour). They've demonstrated a clear relationship between the number and kinds of words children *hear* and the number and kinds children *produce*. In a nutshell, children who hear more language and more complex language in everyday interactions tend to produce more language themselves. What seems to matter is *not* what children are explicitly taught about language (for example, through the warnings that many of us heard growing up, such as "Don't end your sentences with a preposition"). Rather, what counts is what children hear and are exposed to in their day-to-day lives through everyday conversations, at the dinner table, in the car on the way to the grocery store, or in the backyard.

QUICK TIP: Immerse Your Child in Language All Day

Children learn their first, second, and third languages best by being exposed to rich, dynamic, engaging interaction in each of those languages. The best way to get your children talking is to surround them with language.

First language learning happens as part of routine interactions in everyday life. There is no explicit teaching here; for instance, to our knowledge no community of parents ever explicitly teaches their young to conjugate verbs in the past tense. Rather, first language learning is pleasurable, intimate, and interwoven with everyday life in meaningful ways. To be optimally successful, learning a *second* language should be much the same. In other words, second language activities—with older and younger children alike—are most effective when they are fun, playful, interactive, and connected to everyday activities. This sounds good, but what does it mean in real life (and in real homes, where there's laundry to do, carpools to coordinate, dinner to cook, and lots of other immediate demands)?

HOW TO BEST SUPPORT SECOND LANGUAGE LEARNING IN YOUR HOME

Parents who want to promote a second language in their home often feel they don't have time. Modern life—and modern parenting—are both incredibly busy, and learning a second language can seem to be an extra add-on to already overscheduled lives. There's good news for the busy (and the weary) here, however. These two realities—first, that language learning happens naturally in real-life contexts, and second, that day-to-day living is often too full of activities and obligations to squeeze anything else in—actually *complement*, rather than compete with each other. Second language learning works best when it is integrated into all of our busy everyday routines and many activities. It doesn't need to be an extra.

Not all parents are fluent speakers of the language they hope their children will learn, and so, how child language learning happens—and what's most feasible—will differ by family. We'll consider best strategies here for three family language profiles: *majority*, *mixed*, and *minority*. While these are broad-stroke categorizations, parents usually find it helpful to identify which profile fits them best and to consider possibilities accordingly. At the end of the chapter, we'll discuss in detail suggestions concerning babysitters, playgroups, and grandparents that are relevant for *all* families.

If You Are a Majority Language Family...

In *majority language families*, parents are monolingual speakers of the majority language. For instance, in the United States, Britain, New Zealand, and Australia, this means that the parents speak English as their first and best language. In Mexico and Spain, it means that the parents are native Spanish speakers. Majority language parents might have some limited knowledge of a second or foreign language, or they might be completely monolingual. These families generally are not worried about teaching or maintaining the majority language (nor should they be) so much as introducing a second language to their children at a young age.

For instance, Alison's family is a majority language family. She was born and mostly educated in Britain. Her husband Dave was born and educated in the United States. They speak English with each other (of course) and each has a smattering of other languages. So, Alison knows a little Japanese from when she taught English as a Second Language in Japan. Her husband learned Spanish in high school and college, and Japanese during a study abroad program.

They are trying to introduce their two-year-old daughter Miranda to both Japanese and Spanish, as well as keeping her proficient in two dialects of English (British and American). Alison makes sure Miranda interacts with plenty of people who speak British English, and both Dave and Alison involve her in as many Japanese and Spanish language–based activities as possible. She has music and movement classes in Spanish, a Spanish lesson once a week at her preschool, and one of her babysitters speaks Spanish. At home, they have a weekly Japanese night when Miranda dresses up in her Japanese clothes, they eat Japanese food, and they try to speak only in Japanese. To supplement their limited language skills, they also have a Japanese-speaking language student come over and play with Miranda during the semester. When she is older, Miranda will enroll in formal Japanese language classes. So, within her majority language family, Miranda is growing up understanding that there is more than one language with which to express herself. As she gets older she may keep up with both Spanish and Japanese, or she may settle on one.

For majority language families, use of the second language can (and in optimal circumstances, should) begin in the first months of infancy. Parents should try to incorporate the second language into as many aspects of life as possible. Even parents who know only bits and pieces of the second language can use it with their infants and babies. Young babies love the sound of human voices and one language is just as good as the next for soothing, entertaining, or stimulating them. Mothers or fathers, for instance, can sing the same nursery song in the same second language over and over (after all, babies love repetition!). One mother we know sang "*Los Pollitos*" (a popular Spanish-language children's song about baby chicks) in Spanish every single time she changed her son's diaper. He loved it—and was so entranced that he didn't wriggle around during the changing routine—and before she could believe it he was singing it with her. Another father we know counted in Spanish each and every time he went up and down the stairs with his young son. This sounds very simple, but allowed two-month-old Jacob to regularly hear simplified, but real Spanish many times a day in a way that was connected meaningfully to real life (in this case, to real motion, as he felt each and every step up and down in his father's arms!). Another mother we've worked with made a point of learning the phrases for simple games such as "where's the baby?" and peekaboo in French so that she could always (and only) play these games in French with her daughter. She herself was surprised to hear her eighteen-month-old daughter say *"Où est le bébé?"* ("Where's the baby?") to her baby doll one day!

Majority language parents should not worry too much about speaking the language absolutely perfectly or with a non-nativelike accent. It's less important for your child to hear, for instance, perfect Korean than it is to have some early and meaningful exposure to Korean. And there's lots of evidence that language learners can benefit greatly from interaction in the second language, even if it's not coming from the mouth of a native speaker of that language!

In each of these cases, parents are not doing anything extra. They are just doing the routine talk of everyday life, but doing it in the second language. This approach has a number of advantages. Perhaps most obvious is that it doesn't require much extra work for the parents—other than memorizing some key phrases or silly songs of course. Most parents talk to their children much of the time; the only difference is that parents are doing that talk in a different language. An added bonus: New parents often complain of feeling a bit brain dead and in need of mental stimulation in those very time and labor intensive (and repetitive) first months—this approach provides a perfect way to make those everyday routines a bit more challenging and stimulating for mom, dad, or whoever is the caretaker. Parents might even find that they will brush up on their second language skills, too!

Aside from being relatively simple (and potentially entertaining) for the parent, this approach also corresponds with what we know about child language learning. In particular, we know that children benefit from live, human interaction. Infants in particular enjoy and benefit from repetition, repetition, and yet more repetition. Further, these sorts of games and rituals correspond to what we know about language learning more generally—that is, it is more effective when it is fun, integrated into real life, and meaningful for the child.

The options and opportunities for language learning and interaction become more plentiful as the child grows (and as the parents recover from the first few months of sleep deprivation and total exhaustion!). For instance, even parents with only very basic proficiency levels in the second language can read simple board books with their children in that language. These books can be read over and over (and over and over). Young children can learn basic vocabulary in this way, become familiar with different styles of language (for example, formal versus informal), and of course develop a lifelong love of reading.

SPOTLIGHT ON RESEARCH:

Reading Books Promotes Language Learning!

Researchers High, LaGasse, Becker, Ahlgren, and Gardner conducted a study with five- to eleven-month-old children from 205 families in 2000. The parents were recruited during their visits to the pediatrician's office and were interviewed about their home reading practices. The families were then randomly divided into two groups: (a) an intervention group that received a book at the time of each visit from the pediatrician and a handout on the benefits of reading to young children and (b) a control group that did not receive these materials. After three visits (or when the child was twenty-two months old), the researchers interviewed the parents again to see if their reading practices had changed, and they then assessed how many and what types of words the children knew. Parents who received the books and information did report increased reading with their children as compared to the control group. They also found that for older toddlers (eighteen to twenty-five months), this increase in reading aloud was linked to a higher score on vocabulary skills relative to same-age toddlers who did not receive the intervention. In a nutshell, reading aloud clearly seemed to affect these toddlers' language skills.

Recent research indicates that children's vocabulary size in their first language is directly related to how often they are read to in their first language. What about when there are two languages in the picture? In 2002, Janet Patterson found that the same held true for children learning two languages. She looked at the spoken vocabularies of sixty-four two-year-olds who were learning Spanish and English. She found that the size of their vocabularies in each language could be predicted by how often they were read to in each language. In fact, the frequency with which children were read to in a foreign language had more of an impact than even the total exposure they had to the language. Reading everyday children's books is a great way to teach children vocabulary words, and Patterson's work suggests that reading may be a better vocabulary booster than conversation in that language. This is something to keep in mind each and every day. (See the back of the book for suggestions on great places to buy foreign language books.)

In sum, language majority families will likely want to strive to incorporate the second language into everyday routines and activities as early and as much as possible. Parents can do this in fun, interactive, and meaningful ways, not by explicitly teaching the language, but by integrating it into everyday life. Alison's family, as we mentioned above, does this in three main ways: attending classes taught in Spanish in the community, organizing Japanese nights at home, and hiring babysitters with language skills. This is less work and more fun than most parents had ever imagined second language instruction could be!

POINTS TO REMEMBER FOR LANGUAGE MAJORITY FAMILIES:

- It is important to incorporate the second language into everyday routines and activities as early and as much as possible. This can be done in fun, interactive, and meaningful ways.
- Even parents who know just a little of the second language can incorporate it into silly songs, games, and other intimate routines.
- Consider having a fun family night that centers around the second language. This can be as simple as making (or ordering) a pizza and talking about all the ingredients in Italian.
- Reading to children in a second language is one of the best ways to boost their vocabulary.

If You Are a Mixed Language Family...

In *mixed language families*, at least one parent is proficient in a language other than the majority language. This may be because the parents have different language and culture backgrounds. So in many mixed families, parents have two different first languages. However, in some mixed families, both parents have the same first language, but one of the parents has acquired a very high level of competency in a second language. Parents in linguistically mixed homes tend to be concerned about promoting development of both languages, and often worry (with good reason) that the majority language will become much stronger.

For instance, Eury and Thorsten have a five-year-old daughter, Sarah. Eury speaks only English. Thorsten is from Germany and speaks German as his first language and English as his second language. He came to the United States eight years ago as a young adult. Eury and Thorsten have tried to provide opportunities for Sarah to learn German and experience German cultural activities.

Thorsten mostly speaks German to Sarah, and they sometimes watch cartoons on the Internet in German. She attends a German class at a local community center once a week. However, despite all this, Sarah speaks to both of her parents in English, and in the summer, when she doesn't go to her German class where she has to speak German, she often goes weeks without speaking a word of German. Even when her German grandparents come to stay, because of their high levels of English, Sarah is able to speak English with them. Despite all this, her parents keep trying and believing that one day, perhaps in high school, she will be glad for all the German exposure she has received.

For mixed language families like Sarah's, many of the same principles and practices generally apply, most importantly that both languages be incorporated into many aspects of everyday life. However, mixed language families also have additional factors to consider, perhaps the most important of which is finding the right balance between both languages. In most mixed language families, the aim is for children to have roughly equal competence in their languages. In striving to meet this (ambitious) aim, it is crucial that parents consider (1) the *quantity* and (2) the *quality* of input and interaction in both of those languages.

By *quantity* of input, we mean how much of each language does the child hear each day, and in particular, how much of each language is directed at the child. How much input is needed? While there aren't tons of data out there on this question, what we do know fits with general intuition and common sense. For instance, some research suggests that in order for children to have the best chance at roughly equal levels of competence in both languages, they need roughly equal levels of exposure to both languages.

SPOTLIGHT ON RESEARCH:

How Much Is Enough?

In 1997, Pearson, Fernández, Lewedeg, and Oller studied twenty-five Spanish-English bilingual children who had varying levels of exposure to each language, some with two bilingual parents, others with only one parent who was proficient in one of the languages, and others with only a nanny or babysitter as the source of Spanish. They found that the vocabulary learned by children was proportional to how much they had been exposed to each language. However, with as little as 20 percent of total exposure in one language, children were still able to develop a productive vocabulary. In other words, their findings suggest that children could actively speak their second language if they were exposed to it for one-fifth of their waking hours. On the other hand, if total exposure was less than this, children were very hesitant to speak in the language at all. So, if active use of two languages is the goal, for a two-year-old child who is awake about twelve hours each day, at least about two and a half of those hours should be spent interacting in the second language.

By *quality* of input, we mean what sorts of interactions your child engages in with language. For instance, does your child have the opportunity to hear songs and play games in both languages? Are there equal opportunities to interact with books and written materials in both languages? In other words, does your child have the possibility and incentive to interact in a range of different ways (playing, singing, being read to) in a range of different contexts (at the zoo, at the grocery store, in the kitchen) in both languages? These aren't easy questions to answer. Mixed language family parents should consider conducting a Family Language Audit, which will help them consider carefully how much of what types of language learning opportunities exist for the child. (No professional accountants needed here! Tips and tools for doing a Family Language Audit for your own family are at the end of this chapter.)

One-Parent–One-Language in Mixed Families

In many mixed language families, parents opt for what is known as the one-parent–one-language approach, where each parent speaks a different language (typically their mother tongue) to the child. This approach is popular in part because it can help ensure roughly equal language exposure for the child in terms of both quantity and quality. This is assuming of course that both parents share roughly equal responsibility for child care. A side note that drives home this point: We have worked with many families who wonder and worry why their children are not using the father's language. After the family conducts a Language Audit it sometimes becomes clear that the father only engages in intensive *solo* child care for three or four hours total per week. As this is fewer than 5 percent of the children's waking hours, it's hardly surprising that they don't speak the father's language!

The one-parent–one-language approach has many advantages for both parents and children. It is relatively straightforward and does not involve complicated rules or systems (such as only French on Fridays or the Japanese-night scheme used in Alison's family). Typically, each parent speaks the language he or she knows best with the child, so to many parents this approach seems the most natural. Children learn to associate one language with each parent right from the start and, *ideally*, start to use that language with each parent accordingly as they begin to talk.

For these reasons, the one-parent–one-language approach is often held as the gold standard in bilingual child rearing. Yet while there are many successful outcomes (that is, adult bilinguals from such families), it's worth noting that there are also many children reared in one-parent–one-language families who become passive bilinguals rather than active ones. That is, they can understand both languages very well but only speak one of the languages proficiently. This is exactly what was happening with Eury and Thorsten's daughter, Sarah, who understood German perfectly well, but insisted on speaking English. We've even met families where *both* parents speak, say, Swedish, to the child, but the child decides from the get-go that English is her language. Why does this happen and what can parents do?

The Hot-House Approach

One of the reasons for these varied language outcomes in terms of children's competencies and preferences has little to do with what's happening *inside* the home, but everything to do with what is happening *outside* it. Even very young children are sensitive to what language is spoken around them—in the sandbox at the park, at the supermarket and dry cleaner, by their older cousins, and so on. In nearly every community, one language is dominant, and children very quickly pick up on this fact and gravitate toward it like a high-powered magnet. And children grasp this—often before they themselves can talk. This is something we've heard again and again from parents we've worked with over the years, and which is also supported by extensive research. Just as, like it or not, most children can identify the golden arches before they can pronounce Happy Meal, they *get* what language is cool.

Given that two languages are never on exactly equal footing *outside* the home, parents in mixed families should consider providing extra support for the weaker, minority language *inside* the home. This is sometimes known as the

> **• FAST FACT •**
>
> Protecting the minority language at home by using it *more* often, say 80 percent of the time, than the majority language can give children a better shot at becoming active rather than passive bilinguals.

hot house approach, where the weaker language is safely nurtured in the home in the early years. The idea is that after several years in hot house conditions, the child's language competence will be strong enough to withstand extensive and intensive exposure to the majority language, for instance, when they begin formal schooling. In practical terms, this might mean that if you are a parent in a mixed language family, you might want to set a goal of 80 percent child interaction in the weaker, minority language and 20 percent in the majority language. Such a ratio has the advantage of providing a strong foundation in the first language, but also providing for some second language learning early on as well. See more tips on how to deal with this prestige imbalance in chapter 12.

POINTS TO REMEMBER FOR MIXED FAMILIES:

- It is critical for parents to provide conditions for the right balance between the two languages.
- Conducting a Language Audit can help you become more aware of language patterns in the home (and future language competencies).
- Mixed families might want to consider implementing the one-parent–one-language or even the hot house approach, depending on their circumstances.
- Language learning, whether first or second, is most effective when it is enjoyable, integrated into everyday routines and interactions, and meaningfully connected with real life.

If You Are a Minority Language Family...

In *minority language families*, both parents are native speakers of a language that is not the majority language of the region. So for instance, these families would include Polish-speaking parents in

London, Taiwanese-speaking families in Vancouver, or Spanish-speaking Peruvian parents in Las Vegas. Minority language parents often worry about their children maintaining knowledge of their language—Polish, Taiwanese, or Spanish, for instance—but also about them learning the dominant, majority language as well (for example, English).

Our Korean friends Seok and Hye-Jeo and their son, Nathan, provide an example of this sort of family. They came to the United States to attend graduate school fifteen years ago. They graduated, became professors, and decided to stay. Seok's mother also immigrated to the United States from Korea following the birth of Nathan, and she lives with the family. They live in a vibrant area of Philadelphia, which is home to many other first- and second-generation Korean families. However, since he was four, Nathan has attended a private school where English is the medium of instruction, and not many other children speak Korean at home. Seok and Hye-Jeo only speak Korean at home, and Nathan's grandmother speaks no English. They went through a brief period of worry and doubt when Nathan first started school, and his teacher reported he was not speaking very much in class and seemed very shy. However, by December, Nathan had made many friends and began to have playdates and has since thrived. Now Seok and Hye-Jeo are starting to worry not about Nathan's English, but rather, how to help him maintain his Korean.

Language minority families, where both parents speak the minority language as their primary language, will also want to consider these outside-the-home issues in thinking about the inside-the-home patterns. Many minority language parents share twin concerns: (1) that their children learn and maintain the minority language and (2) that they also master the majority or dominant language. As in the case of Seok and Hye-Jeo, an all-English school environment together with consistent use of Korean at home helped Nathan become bilingual (although they recognize that English will likely play a bigger role in Nathan's life as he grows). But what about families who are afraid this won't be enough? We are lucky to have lots of research on these goals, as well as on how they are interrelated and actually support each other.

For parents who want their children to learn and maintain their heritage language, the very best and most important thing they can do is *speak it with them regularly, consistently, and naturally*. How well children speak is directly related to how much they are spoken to. For a shot at something approaching balanced bilingualism (that is, equal levels of competence in both languages), the minority or heritage language must be the primary language used for communication between at least one parent and the child. The language does not need to be taught in any formal or scripted way (in fact, we recommend against this for younger children); it does, however, need to be the main, everyday, and routine language of the home and family connections.

SPOTLIGHT ON RESEARCH:

The Magnetlike Pull Toward English

Maintaining even a big language like Spanish in the United States can be a bit like swimming against the tide. Recent research shows how within a few generations after families move to the United States from Latin American countries, fluency in Spanish all but disappears. In 2006, researchers Rumbaut, Massey, and Bean examined demographic data from Southern California's large Latino population. Their findings suggest that Mexican immigrants arriving in Southern California today can expect only one in twenty of their great-grandchildren to speak fluent Spanish. As they explain, "Even in the nation's largest Spanish-speaking enclave, within a border region that historically belonged to Mexico, Spanish appears to be well on the way to a natural death by the third generation of U.S. residence."

While Spanish maintenance among the descendants of Mexican

and Central American immigrants was slightly higher than among other groups, they still followed the usual pattern of switching to English as the years passed. For instance, among Mexican Americans with two U.S.-born parents (but three or more foreign-born grandparents), only 17 percent spoke fluent Spanish. Among those with only one or two foreign-born grandparents, Spanish fluency was only 7 percent. And only 5 percent of Mexican Americans with U.S.-born parents and U.S.-born grandparents spoke Spanish fluently. As they observe, their findings illustrate clearly that English is not threatened: "What is endangered instead is the survival of the non-English languages that immigrants bring with them to the United States." This data is also a good reminder that all parents—even those fluent in a language other than English—need to be vigilant about promoting bilingualism.

But what about learning English (or whatever the majority language of the community may be)? Many minority language parents are understandably anxious to promote the majority language early on. After all, this is normally the language of school and success. While this fact is undeniable, we are continuously surprised and dismayed to hear from minority language parents from around the United States that they have been advised by their children's doctors, teachers, and speech therapists to speak *only* the majority language, that is, English, with their children in order to avoid confusion or delay. There is *no basis* for these claims, and indeed, what research exists in this area in fact suggests the opposite: Bilingual children have greater knowledge about how language works as a system. (More on how to handle such misguided recommendations in chapter 10.)

There is abundant research that a strong first language lays the groundwork for a strong second language; the two languages support rather than undermine each other. By using their first language at home, parents are supporting the future development of the second language (that is, English). Although this sounds counterintuitive, there is actually a large body of research support for this basic principle. For example, Skutnabb-Kangas and Touko-maa studied the role of the first language (Finnish) in learning the second language (Swedish) among Finnish immigrant schoolchildren living in Sweden. They found that the children who were most successful in learning Swedish already had a high-level mastery of their first language, Finnish. Those without strong skills in their first language (Finnish) struggled most with Swedish. In short, strong skills in the first language *help* not hinder acquiring them in a second.

 SPOTLIGHT ON RESEARCH:

Comparing U.S. Latino Children from English-Speaking and Spanish-Speaking Homes

In 1985, researcher David Dolson set out to investigate what happened to Latino children when their Spanish-speaking parents switched to using mostly English at home. He recruited 108 children for his study: About half came from families where Spanish was maintained as the main home language and about half were from families that had switched to mostly English. The families were similar in other important ways, such as their economic status and the length of time that the children had been in school.

Dolson found that the two groups performed similarly in terms of (1) English reading skills, (2) oral reading skills, (3) length of time classified as Limited English Proficient by the school, (4) attendance rates, and (5) disciplinary referrals. He found that the children from Spanish-speaking homes performed significantly better on measures of (1) mathematics skills, (2) Spanish reading vocabulary, (3) academic grade point average, (4) effort grade point average, and (5) retention. This work strongly suggests that there are important advantages (and no apparent disadvantages) to continuing to use Spanish at home!

Other research has examined how language choices in minority language homes relate to how children perform on academic tests. For instance, Catherine Snow, a Harvard researcher, worked with children at the United Nations International School in New York City, where many of the students lived in homes where languages other than English were spoken (the school primarily serves U.N. employees). She wanted to know if (and how) home language use patterns were linked to academic performance. She administered a battery of word definition and standardized tests to three groups of students: (1) those from English-speaking homes where only English was spoken; (2) those from minority language homes where the minority language was used regularly with the children; and (3) those from minority language homes where parents used English with the children. She found that children who grew up in monolingual homes where the language of communication was not English fared just as well as monolingual English-speaking children on the tests. In other words, scores from groups 1 and 2 were very similar. However, children whose parents spoke their non-native language, English, at home (group 3) fared considerably worse. As an explanation, Snow suggests that the parents' "use of their native language was beneficial because it provided rich, complex language input for the children."

To sum up, language minority families will also want to consider carefully their home language patterns. They should feel comfortable surrounding their children in their first language from infancy. If parents aim for their children to learn their own language, it can and, in most cases, *should* be the principal language for parent-child communication. The second language can be introduced at a young age as well. Language minority parents need not worry about confusion, but should monitor

• **FAST FACT** •

A strong foundation in the first language is linked to successful learning of a second language as well as academic achievement.

closely the quantity and quality of the child's exposure to each of the two languages, bearing in mind that the majority language has a way of seeping into each and every household. For this reason, they might want to do the Family Language Audit at the end of this chapter.

POINTS TO REMEMBER FOR MINORITY LANGUAGE FAMILIES:

• Consider using the minority language at home more often than the majority language in order to provide space for that language to grow.
• Keep an eye on both quantity and quality of language exposure.
• Remember that the majority language will tend to seep into many family contexts, and it's easy for this to go unchecked.
• Using your native language with your children is the best way to ensure they'll be bilingual.
• Using the language *you* are most comfortable with when talking to your children is important for their cognitive and academic growth.

SUPPORT FOR PROMOTING LANGUAGE LEARNING IN THE HOME

No matter what your family type, chances are very good that your home language learning efforts could get a big boost by strategically making use of babysitters, playgroups, and extended family, and by doing a Language Audit. These powerful resources are relevant for *all* families.

Babysitters, Nannies, and Au Pairs

Language learning at home can be significantly supported through strategic use of caregivers, including babysitters, nannies, and au pairs. This is because language learning happens best through natural, everyday interactions. Parents are not so much doing something extra here, but doing what they normally would (that is, finding child care for their child), and doing this with language learning as an additional goal. It is important that the child-care provider be treated as a professional and that parents are explicit about the language and types of language they expect the caretaker to use with the child. Caretakers—like everyone else—are happiest and do best when their expertise is valued and explicitly noted.

So what are your options for child-care providers and how can they support language learning in the home? Here's a brief rundown:

Babysitters, as parents know, are typically hired by the hour on an as-needed basis. Babysitters are often students or neighborhood teenagers, but sometimes older adults as well. Most babysitters are found through word of mouth and friendship networks. It's also possible to find babysitters through community bulletin boards (both physical ones and online) and high school and college employment offices. When posting an advertisement for a babysitter, consider adding a note that "proficiency in X language desirable." And when interviewing potential babysitters, important questions to add to the list are what languages does she know and how well. Once you find a babysitter who meets all of your needs, it is important that you talk with her about language along with all the other routines (for example, emergency numbers, bedtimes, suggestions for dinner). Because language learning happens best when it is embedded in a fun and meaningful activity, it may also work well to suggest specific activities that the sitter and child(ren) can do together. These don't have to be complicated or expensive, but they should provide lots of opportunities for interaction. For instance, arts and crafts activities (like making model clay figures or finger painting) or trips to the local ice cream shop or the library (especially if they can read and select books in the target language) provide more opportunities for talk than watching a movie or going to the playground.

Nannies are typically hired as full-time child-care providers, sometimes living with the family they work for. Many nannies have devoted their professional lives to child care, and tend to have the most experience (and also to be the most costly as a result).

SPOTLIGHT ON RESEARCH:

Nannies Can Be Excellent Language Teachers!

As research on language acquisition has tended to focus disproportionately on child language development in the context of a traditional nuclear home (mother-father-baby), there has been little research on the role of other caretakers (like nannies) and to what extent nannies can effectively serve as language teachers for children. To investigate, Kendall King and graduate students Lyn Fogle and Aubrey Logan-Terry followed a small number of families, each with one young child between the age of one and two over the period of one year. Over the course of the year, each family conducted monthly recordings in the home of the children and their different caretakers doing routine activities (for example, playing with toys in the living room, eating dinner, having a bath). We looked closely at the quality and quantity of language used by caretakers, and in particular compared that used by the mothers and the nannies. We found that overall nannies talked *more* to the children than the mothers (as measured by the total number of words per session). Their language was equally complex (as measured by the range of vocabulary and the length of their sentences). And perhaps most interestingly, the nannies were better at sticking to or enforcing the target language than were the mothers. So for instance, in one family, when the child used an English word for *train*, the mother would confirm *train* in English, while the nanny in that same family would respond with "*sí, un tren.*" While both the mothers and the nannies were bilingual, the nannies tended to stick to Spanish. Although these findings are based on only a small number of families and cannot be generalized to all caretakers, they do suggest that nannies are potentially excellent language teachers. Why? None of these nannies were ever trained as language teachers, but they consistently used the language in fun and meaningful ways with their charges, and language learning was integrated into all aspects of the child's life.

Au pairs are most often young Europeans who live with families for a set period of time, usually a year. Au pairs typically work full-time and are treated a bit like members of the family (for instance, often preparing and eating dinner together). Au pairs often have good English skills that they are interested in polishing even more, and may need the most encouragement to use their native language with the children. As with other caretakers, it's important to talk explicitly with au pairs about your hopes and expectations concerning language. It is also beneficial to establish some regular language-rich activities that au pairs and children might enjoy together. In addition, it may be helpful to establish some language rules in the house that give the au pair the chance to establish a role as the language expert. For instance, the whole family might try to talk in her language three nights a week at dinner, or all of Friday could be language immersion day for the whole family. Lastly, be sure to talk to her about *the au pair's* language goals to make sure you are working together to meet everyone's aims. For instance, if she is hoping to perfect her English, perhaps you could help her find a community English class or make sure to speak to her in English once the little ones are in bed.

QUICK TIPS: Talking to Your Child-Care Provider About Language

- Let her know that you value her particular language skills. ("We are thrilled to have found someone who is a fluent speaker of Polish to work with our daughter. We see this as an important asset.")
- State explicitly that you hope and expect she will use her native language with your child to the fullest extent possible. ("The fact that you are a native speaker is one of the reasons we thought you were a good fit with our family. We hope you'll use Polish nearly always in the house.")
- Be specific that you do not want the language to be formally taught. In other words, you do not want or expect "language lessons" on colors, numbers, body parts each day. ("We really believe that the best way to learn is through natural interaction, just always using Polish. Don't feel like you need to prepare anything formal to teach her.")
- Let her know that you hope she'll immerse your child in the target language by using it naturally, regularly, and in meaningful ways throughout the day (for example, while feeding, bathing, and changing your child). ("It would be great for Anita to hear Polish during all her activities: playing games, cleaning up, eating, and even taking a bath. Just use it naturally!")
- Reassure her that your child will quickly catch on if she keeps things clear and simple at first and that it is okay with you if the child initially doesn't understand everything. ("I know that Anita will not understand everything at first, but we're completely confident that she will learn quickly. Just continue to speak to her in Polish, and it will come in time.")
- Ask her opinion and check in with her regularly about how things are going in terms of language and communication. ("How is your time together going? Do you have any suggestions for how we can help Anita? Is there anything you are concerned about?")

• Support her and your language learning efforts by leaving out children's books, adult magazines, etc. in the target language. ("I left some new Polish books on the table if you have time for reading today! Help yourself to the *Time* or *People* magazines while Anita is taking her nap.")

Playgroups

One of the best ways we know to promote early second language learning is to form a language-based playgroup. Playgroups of all sizes and types have many, many advantages. Our top five reasons why they are worth the effort:

• *Playgroups provide a free or inexpensive way to meet and share tips with likeminded others.* Where better to share your enthusiasm, your doubts, and your hopes, while also learning about the local bookstore's toddler hour for Spanish book reading? Playgroups are for kids of course, but they are also for parents, providing camaraderie, support, and a friendly ear.

• *Playgroups provide parents with real-life interaction in the second language being used with other children in natural ways.* Many of us learned our second language formally in school, and if we are trying to use our second language with our kids, we can feel a bit awkward with our baby talk. Seeing other parents use the second language can be reassuring and helpful.

• *Playgroups give your children regular exposure to the language.* What better motivator for your child than to see a slightly older one model a key phrase like "that's cool!" or "it's mine!" in the second language!

• *Playgroups can be geared and organized to meet the needs and schedules of different families.* Since the parents are in charge, they can set the times, locations, and activities to their needs and desires.

Assuming we've convinced you that playgroups are an excellent way to support language learning in the home, your next question might be how to find or form one. Of course, checking to see if one exists in your community or neighborhood is the first and easiest step. Sometimes these are posted on community bulletin boards (say, at the grocery store, local coffee shop, or library, or in the local newsletter), but in our experience, the best way to find them is through word of mouth. Ask around at the parks, swimming pools, and playgrounds. The turnover rate in playgroups tends to be high and only rarely do all members show up for a particular meeting date, so the best ones are often flexible, welcoming new members on an ongoing basis. If you don't find one that fits your needs, never fear, it's very easy to form your own!

The first step to forming a new playgroup is to find a few like-minded parents who are keen to promote the same language with their children of approximately the same age (and who share roughly similar schedules and time preferences). However, we don't mean that parents themselves need to have the same language proficiencies; we've seen many successful playgroups, for instance, in which about half of the adult members speak Spanish as a first language and about half are learning Spanish as a second language.

 QUICK TIP: Forming a New Language-Based Playgroup: Sample Announcement

Attention Czech Speakers! Mluvíte česky?
Are you finding it difficult to find Czech-speaking friends and play-mates for your children? We are a group of parents looking for other parents and kids to join in a Czech-speaking playgroup. We meet every Wednesday at 10:00 a.m. for one hour of songs, play, conversation, and fun in the Czech language. If interested, please contact Petra for more information: 555–1234.

It's best to first find a core group of members, typically one or two other parents who are very interested and motivated. This is best done by asking your friends, family, and community networks. If you don't find a core group in this way, don't fear, move straight ahead to step two: advertising your group. This is much quicker and easier than it sounds. We suggest writing a simple description of the group and contact name and then posting this: (1) around your community on local bulletin boards or in the local paper (just keep a stack with you and put them up as you go about your everyday errands); and (2) on local parenting list-serves and e-mail groups if they exist in your area.

Playgroups should have no more than fifteen members, maximum. (Although in our experience usually fewer than half of the group turns up for any particular meeting—still fifteen adults and kids is probably the very most anyone would want in their living room on a rainy afternoon!) Once the group gets beyond this number, it should consider splitting into two different groups or possibly looking for a community center, church basement, or other low- or no-cost locale for meetings.

It's important to establish some working rules for the playgroups, and it's often a good idea for there to be a loose structure to each session. For instance, in many of the most successful playgroups we've seen, there is a regular meeting time (for example, every Wednesday morning from 9 to 10:15), and the host rotates each week. Playgroups based at home or community centers seem to work better than those at parks or playgrounds, where it is much more difficult to keep the group together.

At the start of the session, there is usually a short welcoming song or activity in the group's focus or target language (this should be quite short for young children, a bit longer for older kids). This is often followed by a free play period or a planned activity, a very simple snack, cleanup time, and a good-bye song. The only real rule (other than no hitting, biting, or throwing toys) is that the adults talk as much as possible in the target language and the kids are encouraged to do so as well. For bilingual groups, it sometimes works for one week to be in one language and the next week in the other language; that way, everyone has a turn being the one who is more proficient, and everyone gets to practice in his or her weaker language. Once the group becomes established, it's also a good idea to take turns with organizers, so for instance, one person leads the group for three months at a time. The leader is in charge of sending out e-mail reminders and recruiting new members as needed.

 POINTS TO REMEMBER:

- Keep groups small (preferably under ten or fifteen members).
- Have a regular meeting time (once a week is ideal).
- Rotate homes, or meet at a community center where you can be sure to keep the group together.
- Open and close with a short song or activity.
- Allow time for free play and maybe a simple, healthy snack.
- Keep it easy: the only rule is that everyone speaks the target language.

We encourage parents to start thinking about language-based playgroups right from the start. It's even possible to try to drum up some interest and a few members in those preparation-for-childbirth classes. The advantages—both for the parents and for the children— are as numerous as the number of potential groups that exist!

Grandparents and Extended Family Members

Extended family members are crucial players in any family language learning project but can be particularly crucial for minority language families. Grandparents often have an especially close relationship to the child and are particularly important to bring on board. Regardless of their language skills, but especially if they are speakers of the target language, it's important that they be included. Grandparents and other family members may initially be reluctant to participate because they have grown up their entire lives hearing and believing some of the language myths discussed in chapter 2. For instance, they may worry that their grandchildren will grow up confused. Grandparents who themselves struggled to learn English might even take pride in the fact that their grandchildren *only* speak English. It's important that these worries and ideas are brought into the open and directly discussed. (Yes, we know, easier said than done in many families.) Extended family members might even enjoy reading through bits and pieces of this book (especially chapter 1 if they remain skeptics!).

All family members have a role to play in any type of family. Grandparents (and other family members) who are monolingual in English can encourage and promote their grandchildren's language learning by appreciating these skills, by providing positive reinforcement, and showing how proud they are that their grandchildren know more than one language (even if they only understand one of them). Grandparents who speak the target language should be encouraged to use that language as much as possible with the children, and to ignore any kinks or quirks in the child's language. As stressed throughout this chapter, this does not mean "teaching" the language of course, but using it regularly and meaningfully in everyday interactions.

 QUICK TIPS: How to Encourage Grandparents' Involvement

While grandparents and grandchildren are often naturally close, bringing specific language learning goals into the relationship can be challenging:

• *Keep grandparents updated on your children's lives.* This is especially important for grandparents who live at a distance. Use telephone calls, letters and cards, photos, e-mails, samples of children's artwork, and so on. It's amazing how helpful the Internet can be in this respect. Parents can post photos on sites like shutterfly.com, and videos on youtube.com and, with less and less technological know-how, can arrange video chats with their extended family. This can help grandparents feel connected even if they are unable to visit often, which will make the times they do visit more meaningful and natural. If possible, these exchanges can take place in their language.

- *Encourage grandparents to babysit, and in general to spend more time together.* Make it a pleasure, not a chore! Instead of asking for dishes to be washed, suggest activities that allow for interaction, like playing games and going to the park. Make sure they have some quiet time together to just talk (and practice the language).

- *If you have more than one child, consider making time for each child to spend alone with their grandparents.* This allows them to build a more meaningful, personal, and intimate relationship, and helps make the most of interaction in the language.

- *Spoil the grandparents, and don't take their time for granted!* Make a good meal, fill the house with their favorite wine or desserts, and make their stay as pleasant as possible. This will encourage them to spend more time with your family.

- *Make sure they know that they are a valued part of your family* and your language learning plans (and not just a backup babysitter!).

Suggestions adapted from Moore, 2004.

YOUR CHILD'S LANGUAGE "MISTAKES": PROBLEMS OR PROGRESS?

One final point, for all children—from *all* three types of families: Mistakes are a natural part of the learning process. They will say cute things like, "I saw mouses today" or "I holded the baby wabbits." For bilingual children, some of these errors may seem to be the result of learning two languages—for example, a German-English bilingual child saying, in English, "We play now this" (an English sentence that seems to follow German word order). But as we discussed in chapter 2, many of these errors can be taken as signs that the child is learning. For instance, saying "holded" instead of "held" shows that the child is gaining an understanding of the regular past tense -ed marking; it's just that she's overgeneralizing the rule (-*ed* = past tense) for the moment to irregular verbs as well. Likewise, when a child says "mouses," it's clear he's learned about making things plural and about the rule for doing this (adding -*s*). He just needs to hear the irregular plural "mice" enough times to realize that there are some exceptions to this rule.

Be amused by these "errors," and maybe write them down in the baby book, but don't worry about them. When children learn to ride bikes, they fall off quite a few times, even as they're getting better at it. When toddlers begin to feed themselves, there's quite a lot of mess and juice on the clothes, the table, and the floor. Likewise, in learning language—first, second, or more—children will make a few mistakes. It's a natural part of the process.

Interestingly, you might have noticed that your children actually start out using irregular past-tense forms correctly, saying things like, "Daddy went." Then your children might apparently regress and start saying things like, "Daddy goed." What's going on? After all, they had it right before! In fact, what you're witnessing here, too, is the emergence of a system. Before children have much of a grasp of how the past tense works in English, they may simply be using the words they've heard other people use (in this case, "went"). But as the pieces of the system start coming together and making patterns in their minds, children start to engage in more creative production, actually generating the forms themselves. So, although "goed" is ungrammatical, it's a sign that a child has learned the rule -*ed* = *past tense* and is simply in the process of fine-tuning it. Researchers sometimes call this U-shaped learning because the child seems to go downhill for a while, with accuracy rates falling before they pick up the correct form again. This is a short-term phase, though. Especially with plenty of input and interaction, children pretty soon figure out when to apply the -*ed* form correctly. The same thing goes for mistakes that seem to reflect the influence of another language. Children need to get enough meaningful input in each language and given that, the details eventually fall into place.

It's completely okay, and even desirable, for parents and other caregivers not to correct these early mistakes. In most cases, it doesn't do much good to explain to a two-year-old or even a four-year-old the rules of word order in English versus German. Always correcting mistakes is likely to damage your child's self-confidence about speaking the language or even create a negative association with the language—the last thing you want to do! Instead, parents are better off engaging their children in everyday conversations, and through these natural interactions, modeling the correct forms simply by speaking normally. You don't need to correct grammar and vocabulary errors as soon as they appear in order to prevent the formation of bad habits. With older children, some correction is helpful, but you know that already from reading the Which Language and When? chapters!

DOING A FAMILY LANGUAGE AUDIT

Consider conducting a Family Language Audit to begin thinking critically about both the quantity and the quality of your child's language exposure. What we *think* is happening in our homes doesn't always correspond to what is *actually* happening. To get an idea of how much exposure and practice your child is getting in each language, it can be useful to keep a record of how often each language is used during the course of a day, and for what sorts of functions.

Using the chart at the end of the chapter, note each of the activities your child engages in each day (and in which language). Also note whether this activity involves interactive or passive use of the language. For instance, watching *Thomas the Train* videos in English requires passive language skills (listening), while playing a game of Chutes and Ladders requires interactive language skills (listening but also talking and moving the pieces around the board). Once complete, tally the number of hours spent in each language and how many of those were interactive and how many were passive. If you have a fairly stable weekly schedule, keeping track of activities each day for one week can give you a good idea of what kind of language input your child is getting. To see how it works, consider this example from a family with an English-speaking mother, a Spanish-speaking father, and an English-speaking nanny:

 EXERCISE:

Sample Family Language Audit

TIME	ACTIVITY	INTERACTIVE/ PASSIVE	PRIMARY LANGUAGE
7:00	Wake up/breakfast/ get dressed with father	Interactive	Spanish
8:00	Watch videos	Passive	Spanish
9:00	Listen to stories at the library	Passive	English
10:00	Walk and talk with nanny on way to park	Interactive	English
11:00	Playgroup in the park	Interactive	Spanish
12:00	Watch TV programs in Spanish while eating lunch	Passive	Spanish
1:00	Nap	—	—
2:00	Nap	—	—
3:00	Go shopping with nanny	Interactive	English

TIME	ACTIVITY	INTERACTIVE/ PASSIVE	PRIMARY LANGUAGE
4:00	Read books with nanny	Interactive	English
5:00	Play with mother	Interactive	English
6:00	Dinner with mother and father	Interactive	English
7:00	Play in the backyard with parents	Interactive	English
8:00	Bath/story/bedtime with father	Interactive	Spanish

TOTALS:

English: 7	Passive English: 1	Interactive English: 6
Spanish: 5	Passive Spanish: 2	Interactive Spanish: 3

In this example, if you looked only at the total number of hours spent interacting in each language, it seems that the child is getting fairly close to equal amounts of input in Spanish (5 hours) and English (7 hours). However, if you look at the number of hours spent in interaction in each language, twice as much time is spent interacting in English (6 hours) as in Spanish (3 hours). Remember that it is through interaction that children do their most meaningful learning. When taking an inventory of how much input your child is getting, the nature of the activity can be as important as the number of hours.

WRAP UP

Formal education through classes, tutors, and special programs can be an important part of second language learning, and we will discuss this in more detail in chapter 8. However, every parent, no matter what their particular family language profile, can do a lot to promote early language learning at home. The types and amounts of language learning activities you engage in with your child will, of course, depend on your family's language goals and circumstances. For each, however, language learning activities need to be (1) enjoyable and fun for all involved, (2) fully integrated into everyday routines and interactions, and (3) meaningful, interesting, and connected with real life.

It's important to carefully consider your language learning plan as early as possible. This is particularly true if different caretakers will be speaking different languages with your child. Many parents have told us how difficult it is to change the language of communication, between, for instance, the babysitter and the child, or between father and child, once it's been established.

We close this chapter with a reminder that there are sometimes excellent opportunities for inadvertent informal second language learning. By this we mean that in some areas, there are enough speakers of the second language (for example, Spanish speakers in much of the southwest, Chinese speakers in parts of Canada) that there are first language programs established for these groups. For instance, in many Washington, D.C., neighborhoods, there are good independent bookstores that offer story hours each week in Spanish; libraries that provide circle time and stories every other week in Spanish; private schools that offer toddler dance classes in Spanish; and many similar programs.

These provide excellent opportunities for children from minority language families to strengthen their first language skills. They also provide great opportunities for mixed and majority language children to have authentic exposure to the target language. (What's more: many parents report to us that they find it easier and more natural to use the language after having attended, for instance, a Spanish story hour. It keeps them in the groove and gives them something special to talk about with their child.) The key to success here is that these are not only enjoyable and fun for all involved, but that language is being used in meaningful ways. As young children learn the very basics of dance moves, the rhythm of show-and-tell, or of the adventures of Winnie the Pooh, they are also acquiring an additional language.

EXERCISE:

Your Family Language Audit

TIME	ACTIVITY	INTERACTIVE/ PASSIVE	PRIMARY LANGUAGE
6:00			
7:00			
8:00			
9:00			
10:00			
11:00			
12:00			
1:00			
2:00			
3:00			
4:00			
5:00			
6:00			
7:00			
8:00			

TOTALS:

Language 1:   Language 1: (Passive):   Language 1: (Interactive):

Language 2:   Language 2: (Passive):   Language 2: (Interactive):

CHAPTER 7

Language Learning
Through Edutainment:
Can Children Be Educated
and Entertained at the Same Time?

n our fast-paced world, we often rely on technology to do lots of everyday tasks, from checking the weather and reading the news to shopping and paying our bills. And an increasing number of adults are finding they enjoy using technology for second language learning, including software programs like the Rosetta Stone, DVDs of movies in the foreign language they are learning, Web sites, and even Podcasts. Naturally, parents want to know if there are fun forms of so-called edutainment available to help their children learn another language or, if they are already using such things, how well they really work. Of course, the manufacturers claim these edutainment products work wonderfully, as do Web sites that financially gain from advertising them, but what's the real scoop? In this chapter, we provide an unbiased, research-based look at what family entertainment is educational for your child.

For example, some parents have told us that their children love Dora the Explorer, a Spanish-English bilingual cartoon character. Spunky, adventurous, and polite, Dora is a great role model, but parents aren't sure if the limited amount of Spanish actually used on the show is enough to help their children learn the language. Other parents say their children love to sing folk songs from the CD their grandparents sent them and wonder whether to buy more, but aren't sure if their children actually know what words (or even what language!) they are singing.

Here, we review a range of different language learning resources and provide guidelines to help you make informed decisions. Before we get started, though, it might be helpful to reflect on the impressions you've already formed of the different language learning technologies that are out there. What sorts of things (e.g., vocabulary, grammar, accent) do you think children at various ages can learn from each of the following foreign language sources (and how well can they learn them)? Go ahead, fill in the chart!

EXERCISE:

When and What Can Children Learn From?

	In the womb	Ages 0–3	Ages 4–8	Ages 9–12	Age 13 plus
Musical lyrics in the language					
Talking toys and books					
TV/movies without subtitles					
TV/movies with subtitles					
Language learning software					

In the womb	Ages 0–3	Ages 4–8	Ages 9–12	Age 13 plus

Language learning
computer games

Language learning
Web sites

Language
learning DVDs

BABIES AND TELEVISION:
TO TV OR NOT TO TV?

Given all the technological innovations of the last few years, parents now have a huge variety of language learning materials that are aimed at their children. These range from DVDs, computer programs, Web sites, and audio and videotapes, to trilingual talking toys and even dedicated TV channels. In 1997, no one had ever heard of Baby Einstein. By 2003, one of every three children in the United States had watched at least one Baby Einstein video. The market for videos and DVDs for young children is vast, earning about $500 million in 2004 alone. Many of these products are designed and marketed on the premise that children can be entertained, stimulated, and educated all at the same time.

But what does this mean for our children's second language learning? We also wondered. When we first had our children, we too were eager to pop in edutainment DVDs like the *Baby Einstein Language Nursery Video, Brainy Baby*, or a little something from the *Muzzy* series. But after watching a few of these, we started asking ourselves: What types of language learning can take place through these videos? Do children really benefit? How should parents balance any language benefits with the fact that the American Academy of Pediatrics recommends *no television at all* for children under two? So we began to dig a little deeper into the research.

• **FAST FACT** •

WHAT DOES THE AMERICAN ACADEMY OF PEDIATRICS
RECOMMEND FOR TV VIEWING?

Children of all ages are constantly learning new things. The first two years of life are especially important in the growth and development of your child's brain. During this time, children need good, positive interaction with other children and adults. Too much television can negatively affect early brain development. This is especially true at younger ages, when learning to talk and play with others is so important. Until more research is done about the effects of TV on very young children, the American Academy of Pediatrics (AAP) does not recommend television for children age two or younger. For older children, the Academy recommends "no more than one to two hours per day of educational, nonviolent programs."

Source: http://www.aap.org/family/tv1.htm

Other parents, like Molly, also wonder what's best. Molly wants her two-year-old son, Aiden, to learn Spanish—and like many parents, she believes the younger, the better. Molly is a non-native speaker of Spanish. Although she speaks it quite proficiently, she's not fully confident she can teach it to him. Molly wants to help Aiden "get an ear" for the language so he'll have an easier time learning it later on in school. But how? Is it enough to have Aiden watch Spanish-language DVDs and television programs like *Plaza Sésamo* (based on *Sesame Street*)?

To answer this type of question, Patricia Kuhl and her colleagues ran a series of experiments. In a study of nine-month-old babies, one group participated in one-on-one reading and play sessions with native speakers of Mandarin Chinese; a second group watched professionally produced child-oriented videos in Chinese. These videos showed the same native speakers talking in Chinese in an animated and engaging way and playing with toys—just the same as what was happening with the first group of infants, but on the video not in person. The researchers found that the babies who watched and listened to the very high-quality videos did not learn *any* Chinese sounds. In fact, they seemed no better off than a comparison group who had been exposed to *no* Chinese at all. Their findings strongly suggest that for young children to benefit in terms of language learning they need exposure to a living, breathing human who talks and plays and interacts with them in that language.

The upshot of this research—that babies don't learn languages through television—is a bit disappointing both for parents like Molly and for us. Like a lot of parents, we hoped that by providing our children with these DVDs and television, we were not just entertaining them but giving valuable foreign language input. The finding that things don't exactly work this way was a tough pill to swallow and we still wish it were that easy! But the plus side, perhaps, is that we now have even *more* of a reason to engage our young children in interpersonal communications and set them on the road toward active, healthy adult lifestyles. And maybe we don't have to get rid of the TV and DVD player quite yet. Perhaps there is a role for TV and DVDs when the babies become a bit older?

TODDLERS AND TELEVISION

By the age of two, most children watch some television. In fact, the national average in the United States for two-year-olds is two hours of television every day. Given the quality and educational focus of many children's programs today, we might expect (and indeed hope) that toddlers can learn some language—first or second—by watching television.

However, research with lots of different types of children shows very clearly that toddlers pick up almost all of their language by interacting with their parents, caregivers, siblings, and friends. For instance, from research with children exposed to only *one* language, we know that how much television they watch has very little to do with the size of their vocabularies. Children's vocabulary size and use, on the other hand, *is* very closely related to how much their caregivers read to them. Similar evidence comes from studies of the hearing children of Deaf parents (that is, children who are exposed to sign language but to little spoken language in the home). Contrary to the hopes of many Deaf parents (who want their children to become bilingual in sign language and English), hearing children don't pick up spoken language from television. In other words, even though they can hear the stream of sounds coming out of the TV, they learn the language that is used by the real live people around them (sign language). In short, television does not seem to be a good language teacher for either babies or young children. Human interaction seems to feed the brain in a way that TV does not. During interaction, for example, children can test hypotheses by trying something out to see how it sounds and then receive a reaction and some kind of feedback from another person. This sort of conversational give and take is not present with television, yet is extremely helpful for language learning.

But what if your children already have some exposure to two languages? Can television support language learning for them?

> • **FAST FACT** •
>
> Vocabulary is very closely related to how much children's caregivers talk or read to them. Watching TV does not lead to foreign language learning in babies, or increase the size of toddlers' vocabularies.

To answer this question, Janet Patterson studied sixty-four Spanish-English bilingual two-year-olds and their parents. She asked the parents detailed questions about what sorts of activities they did with their children each day in each language. Parents were also asked to estimate how much time their children spent reading and watching TV in each language. Finally, parents were given a vocabulary check-list to mark off what words they had heard their children use. Patterson found that how often the children were *read* to in each language was directly related to their vocabulary size in that language. In contrast, the frequency of watching TV in each language was *not* related to vocabulary size in either language. In a nutshell, reading to children helped them learn new words; watching TV did not.

So why, despite the marketing claims, aren't edutainment devices like TV programs and DVDs doing what we want them to do? As Lise Eliot, a well-known baby expert and neurologist says, "Babies prefer humans over anything inanimate." In her words: "Babies, infants, and children (not to mention adults) learn best from interaction with other humans. It's wired into us. In order to learn, children need language situations where the conversations are interactive, adaptive, and pitched at their level. If the conversations are immediately focused on the things they are interested in, that can only help. This is true for learning in general, and also for language learning."

Given all this, is there a way to support the things we already do (for example, talking, reading, and interacting with our children) with some of the technology that exists for language learning? Let's return to Molly and Aiden to illustrate. When Aiden wants to watch Elmo on a DVD in Spanish, Molly watches it actively with him. For example, she pauses the disc regularly to repeat things, ask him questions, and generally help him talk about the things that are interesting to him. With her level of Spanish, Molly can do this easily. In fact, it's easier for her than if she has to carry the topic of conversation herself, for instance, coming up with each vocabulary word herself. When Aiden wants to watch TV, Molly has a sense of what sorts of Spanish-language programming are available at what times, and she often has the latest *Plaza Sésamo, Dora the Explorer*, or *Go Diego Go* episode recorded so she can put it on and watch it with him.

QUICK TIP: Watching TV and DVDs

DVD recorders and services like TiVo can automatically find and digitally record television programs every time they are on, based on the specified interests of the user. TiVo includes a feature called KidZone, accompanied by lists of TV shows that have been recommended by national parent and children's organizations. Key words like "Chinese" can be used to search for Chinese kid-friendly programs.

What this all means is that technology should not be a centerpiece as much as an accompaniment—something that serves as an object of joint attention. You should generally tune out technology if there are opportunities to talk and interact with your children directly. TV programs and DVDs can be quite useful and stimulating springboards for further interaction, but it is important to keep in mind that it is human interaction itself that is crucial for language development.

Talking Toys and Interactive Books

Talking toys can also provide fun ways for you and your child to use language together. And this is not a stretch from what many parents are already doing in the first place. When Aiden is in the mood to play with his trains, for example, Molly often ends up on the floor talking with him *in Spanish* about trains, tunnels, and tracks. Now, the trains themselves might not be doing all that much to encourage Aiden's use of one language or another, but this may be an area where bilingual toys can play a role, enhancing parents' efforts to engage their children in conversations in another language.

With children like Aiden, who get a big kick out of talking toys, parents may be able to use bilingual toys to set up contexts for speaking the second language. Even if these toys are a bit limited in terms of the flexibility and complexity of ideas they can express, it doesn't mean that a parent can't create creative situations in which the toy (or a magical parental ventriloquist!) is speaking the second language to interact with the child.

QUICK TIPS: Finding Bilingual Toys

Some examples of talking toys: Spanish-English bilingual Elmo sings a song in English or Spanish depending on which hand you press, and he "teaches" Spanish words if you press his tummy. Some Care Bear dolls are programmed with vocabulary for numbers, colors, and common phrases in a second language. Bilingual Language Littles, like Young Hee, are dolls that can recite about twenty-five to thirty phrases in English and Korean if you press one hand, and numbers, colors, days of the week, etc., if you press the other hand. You can find heaps of bilingual talking toys, books, and other games in most toy stores and online.

It can be fun and useful to be able to count, name colors, and say common phrases in another language. But most parents who envision bringing their children up as speakers of a second language are shooting for more sophisticated language abilities. Bilingual dolls, just like bilingual TV programs, should form one part of your much larger tool kit. And again, this tool kit needs to include engaging, meaningful, interpersonal interactions in whichever languages are being learned. Sitting young children down with one of these toys by themselves isn't going to lead automatically to the ability to carry on full-blown creative conversations in the language, of course. However, toys are great for providing opportunities for games and meaningful interactions in a language, and that does lead to language development.

Interactive books are another great way to use technology to support interaction. Books like the LeapFrog series are popular with many parents. The books include games, music, voices, rhymes, and other fun sounds in a number of languages. Some LeapPads set up situations where children can develop their abilities to classify, count, and estimate (for example, by walking them through a shopping trip with the Frog family), or where they can learn vocabulary about household objects, furniture, toys, musical instruments, colors, shapes, emotions, and parts of the body. Since these books exist in bilingual versions, children can do all of this in more than one language. We know of one adoptive father who used LeapFrog games with his Russian-speaking sons (aged seven and eight) to encourage them to build literacy skills in English over the summer before school started. By using technology and games, John was able to promote learning the alphabet and numbers in a fun way without having to put a lot of pressure on his sons as a new father. The LeapFrog activities were just one part of John's approach to language learning as he himself learned Russian well enough to speak it at home with his children and spent a lot of time reading to the boys in both languages. The interactive books gave the boys some fun learning time on their own that added to all the other things they were doing with their father.

POINTS TO REMEMBER FOR TODDLERS:

- Actively watch second language DVDs and TV programs together, pausing, replaying, asking simple questions, and sharing ideas. If you do this from the very beginning, your toddler will expect it and will not see it as an interruption or an intrusion.
- Find second language books about TV characters like Elmo and Big Bird for you and your child to read together and talk about the characters as an alternative to watching TV.
- Pick a word of the week, for instance *pájaro* (bird), after watching a

children's TV show in the language of your choice (for example, there are versions of *Dora the Explorer* in French, Portuguese, Spanish, Japanese, and many other languages). Then use that word to talk about birds, draw birds, and point them out on the street.

• Use bilingual toys and books as a scaffold for creating extensive and meaningful interpersonal interactions in a particular language.

SCHOOL-AGE CHILDREN AND LANGUAGE LEARNING TECHNOLOGIES

School-age children are often very interested in television, video games, and other forms of technology (like cell phones!). Are there times when edutainment *alone* can help a child learn language? Can we as parents *ever* feel like we might be doing a good thing by allowing our children to watch TV or play a language video game after school while we pick up a good book or magazine for ourselves? Happily for all, and especially for busy parents, the answer here seems to be a qualified "yes," particularly when we consider children at or nearing school age.

SPOTLIGHT ON RESEARCH:

Language Learning Through TV in Belgium

Belgium is a small country roughly the size of Maryland with about ten million people. Belgium consists of three separate and relatively independent areas. In one of these areas, Flanders, children speak Dutch at home and learn French in school by age ten. French is the compulsory second language for schoolchildren in Flanders, unlike in other areas of Belgium, where children can choose a second language (often English). So, in Flanders, children don't start learning English, their *third* language, until they are thirteen or fourteen years old. However, when children start their first English classes, they already understand a lot of English words. Why? Because in Belgium about half of all children's TV shows are in English. These shows have Dutch subtitles, but the spoken language is English. And just as for children everywhere, TV and cartoons are a big hit. So, once the children in Flanders learn to read in Dutch, their first language, they are able to use the Dutch subtitles along with the visual cues provided by the action on the screen to help them understand the English that is being spoken. This sort of TV-assisted learning seems to work pretty well for Dutch-speaking children, who arrive at school with stronger English skills than their French-speaking counterparts, for instance, who have much less exposure to English-language television.

In countries like Belgium, children seem to learn some of a foreign language from TV. However, before drawing hard conclusions, it's important to remember that we are talking about hours and hours of exposure over many, many years. And even with all that exposure, children in Flanders are generally not *speaking* English as a result, but only *understanding* it. Still, there are valuable lessons to be learned from the Belgium example. Most importantly, television and other edutainment can serve as important sources of support for a second language and, in particular, they can promote comprehension skills.

What's more: carefully selected TV programs, DVDs, Internet activities, computer programs, games, music programs, and toys can also create positive associations with the language. They show your child that you as a parent value the language yourself. And in some cases, they can expose your child to a variety of real-life situations where the language is used naturally. For instance, our friend Cathrine lives in a small town in the middle of Illinois. She's trying hard to teach her six-year-old son, Magnus, some Swedish, but she's finding it an uphill battle as she has a monolingual English-speaking husband and no other Swedish friends or family in the area. And although Swedish is Cathrine's first language, she's lived in the United States since she was in her early twenties, so she doesn't always find speaking it so natural or easy herself. Cathrine is, however, armed with a DVD player that plays all international formats (available for purchase on the Internet; see the list of resources at the end of the chapter for more information) and a small library of Swedish DVDs (purchased online). As Cathrine explains, "Those DVDs have been a savior for us. They're letting Magnus see other children talking and playing in Swedish, and he's learning new words for things we don't even have here in the United States, like Swedish candy!" Cathrine also points out that videos and tapes of music that she can play in the car give her and Magnus something fun and special to talk about only in Swedish. Magnus also has a cupboard of Swedish toys, including Pippi Longstocking toys, which form another part of his Swedish activities with Cathrine. These, together with a few talking dustpans and brushes (all speaking in Swedish, of course), help Cathrine in her efforts to familiarize Magnus with Swedish and promote positive associations with the language and culture.

Savvy parents are not the only ones who have caught on to the power of language media. Some educators have started joking (or half-joking) about something they call "cybertropism"—that is, analogously to how plants grow toward light sources (phototropism), students grow toward computers. Truth be told, the traditional pencil-and-paper exercises so beloved to us from our youth (or not!) are facing competition from the bells and whistles of various forms of popular media entertainment. While a few children out there enjoy memorizing lists of isolated vocabulary items, many more enjoy watching music videos online, playing video games, text messaging, blogging, wikiing, and Podcasting.

If you have school-age children, there are many ways to harness this technology toward educational goals. One example of a family using technology effectively comes from our friend

Janette and her twelve-year-old daughter, Tanya. Tanya's paternal grandparents speak fluent Russian, but Janette has only a shaky grasp of the language. She wants Tanya to learn the language. It used to frustrate her that Tanya didn't want to study Russian—instead, she just seemed to want to glue herself to the TV after school! But not long ago Janette came up with a great idea: On school nights, Tanya is allowed to watch half of a movie *as long as she plays the soundtrack in Russian.* While Tanya grumbles occasionally about the "Russian on school nights" policy, she also enjoys her movies. It's been so successful that Janette and Tanya now watch Russian movies together on the weekends, making it a special treat for both of them.

Computer Games and Language Learning

Technology-assisted language learning comes in lots of forms. One potential form is interactive computer games like *The Sims* and *The Sims 2.* The language of these games can be changed simply by switching the settings. In *The Sims,* players control the daily routines of virtual people whose personalities and other characteristics they have chosen. As they guide the virtual characters through the processes of finding jobs, decorating their houses, cooking food, entertaining guests, managing finances, and doing a wide range of other activities, the language used is authentic and creative. Parents can even create a bilingual gaming environment in which the stronger language is used to support the weaker one. As with all media, parents need to take care to carefully preview and monitor the content of all games. *The Sims 2* is more targeted Toward teens and adults. A less graphically advanced, but more child-friendly game is

Carmen Sandiego, with the Word Detective and the Where in the World products, with CDs being particularly language relevant.

For children more interested in sports, there is evidence that some language can be learned through, for example, Japanese baseball video games. The graphics in sports games are increasingly true to life, and play-by-play commentators tend to announce the same actions repeatedly using authentic expressions that are very similar to what sports fans hear on TV. Spoken and written material often appear simultaneously, and language use is naturally woven into the playing of the game. As the players' button presses control the moves of the game, they simultaneously control the language of the game.

Online multiplayer games are also very popular with school-age children and can be an excellent way for teenage learners to interact with real, live players (that is, speakers of other languages) from around the world through a virtual community. For example, more than a million and a half people worldwide interact in the virtual community known as Second Life (http://secondlife.com/business education/education.php), where people interact as avatars (or virtual selves, who they have created using tools on the site). A special subsite exists for minors. In the Second Life world, avatars can do anything, including learning a second language (there are classes in a number of languages) and interacting with other avatars using their second languages. Again, as with all Internet activities, parents should always remain vigilant about what their children are doing online.

More Internet Resources

A few other Internet resources worth considering are the same ones that many people encounter every day—for example, e-mail and online news sources. Many e-mail providers, such as AOL, allow users to change their settings so that the advertisements will appear in a different language, and there is certainly no shortage of online media that can be read in languages other than English (just think about how people in Italy or Spain or Japan get their news).

There are also several e-mail- and chat-based initiatives on the

Internet often used by schools to facilitate their classes to engage in interactive communication. Individuals can also search for pen pals in a wide range of language on sites like Linguistic Funland (http://www.tesol.net/penpals/). Blogs can be examined and used, as can video sites like youtube.com, as long as parents are prepared to invest the time selecting the content and often using something like netnanny.com to make sure children stick to what's been selected for them.

Online (and free!) radio stations also abound, with choices available for any taste. The Voice of America Web site lets listeners choose from over fifty languages (it also offers special English newscasts in simplified English, which is much easier to understand for non-native speakers in the early stages of their second language learning), while Live365 provides free music from all over the world. For children who are into watching music videos, written lyrics can be retrieved from online databases and synchronized with songs so that they appear while the videos are playing.

Second language learning materials can also be downloaded from the Internet and then loaded onto cell phones or iPods by adolescents for use anywhere. Online stores like iTunes are a great resource for language learning. You can search for popular artists in relation to their respective languages. The Internet also abounds with Podcasts—free, downloadable audio or video segments that range from amateur productions to slick national radio or TV programs. If your child is learning French and likes fantasy movies, for example, you can look for French Podcasts aimed at Tolkien fans!

Digital media technology can bring language learning seamlessly into everyday life, and school-age children of a range of ages can easily (and enjoyably) take advantage of that. If you're cool enough, they might even let you join them. As with all learning though, parents do need to bear in mind that users of these sorts of media may, because they are viewing it primarily as entertainment, be relatively passive and focus more on reading and listening skills. So, like with all language learning materials, you should be clear about the goals you have in mind (maybe communication strategies in the language, fluency, and vocabulary development, but perhaps not expecting grammar practice). Opportunities for second language learning within the technology world are getting better and better for older children and adults. The number and types are growing every day: See the quick tips, where we share a few of our favorites that you might enjoy exploring with your school-age child.

POINTS TO REMEMBER FOR SCHOOL-AGE CHILDREN:

- If you can afford it, consider purchasing a region-free DVD player to use with your older children, or buy one of the inexpensive software programs that lets you watch all sorts of DVDs on your computer.
- Begin to think about foreign language DVDs (films, cartoons, music) and/or check out your local library. Look for online lenders (like Netflix in the United States or Amazon DVD rentals in the United Kingdom) who have a whole range of international DVDs. Amazon will also let you buy materials directly from any of their sites in different countries.
- Try to watch these in advance to make sure they are age-appropriate, but also to prepare yourself and your child for some of the vocabulary and the story line. Children will enjoy DVDs more (and learn more of the language) if they can follow along with ease!
- Pair DVDs with cultural treats (pictures, food, toys, special games).
- Treat the DVDs as a shared, fun family experience.
- Investigate computer games and online possibilities like video sites.
- Consider buying iPods or other digital media players and loading them with second language learning materials and software for gifts.

WRAP UP

For babies and toddlers, we've seen that for non-native speakers, or parents with even minimal skills in a second language, a few forms of technology can be used to support your efforts to help your baby or toddler learn a language. Some activities include playing games with interactive toys, listening to music, and reading stories or interactive books together. These sorts of joint adult-child activities do much more to promote language learning, and other types of learning as well, than baby videos of any sort. Even though the sellers of edutainment toys for babies and toddlers may want you to believe that they offer the most effective and foolproof means of exposing your young child to a second language, just remember that this is not supported by research.

In contrast, for school-age children, you can feel comfortable about using all kinds of edutainment to support second language learning goals. Television programs, DVDs, Podcasts, language teaching software, video games, and even foreign-language music and music videos can provide positive and fun associations with the language for older children. They can help to link language meaningfully with the culture. And older children, especially those who can read English subtitles or who know a bit of the second language already, do seem to learn some language through these fun and enjoyable combinations of education and entertainment. They can be powerful additions to add to your language toolbox!

EXERCISE:

Brainstorming Ideas for Creating a Positive Learning Environment

Here are a few useful questions to get you started on thinking about whether to use edutainment in your own family (and if so, how):

- Is my child at the right age to benefit from edutainment in any language?
- How can my family make foreign language DVDs/TV an interactive experience?
- What should the ratio of book reading to DVD/TV viewing be in my household?
- What limits or rules should be established for edutainment in my family?
- What resources can I find to help me identify edutainment in the language(s) I'm interested in?

CHAPTER 8

What Are the Characteristics of Good Second Language Learning Programs and Language Teachers and How Can You Find Them?

For lots of different reasons, parents of children learning more than one language often seek language support outside the home. One form this can take is language learning programs at schools or other institutions. Another form is private tutoring. However, within each of these general categories is a sometimes bewildering array of options. Faced with such a range of choices, parents often ask, "What are the key features of a good language learning program?" or "How can I go about finding good programs and good teachers?"

One increasingly popular option is called dual-language (also known as two-way). In most dual-language schools, half of the students are dominant in one language (say, Spanish) and half dominant in another (English). The goal is for students to learn each others' languages and to become bilingual and biliterate in each other's languages after a few years. To many people, this sounds ideal and it's certainly the kind of program Kendall will *try* to get her son into in Washington, D.C. However, two-way immersion is not available everywhere yet (even though program numbers are increasing dramatically), and they can be difficult to get into because of the very high demand for them. What's more, as with most of the other topics we've covered in this book, these sorts of programs are not one size fits all, and certainly not for everyone. In this chapter, we discuss programs like these, and we also provide some helpful tips for parents like Alison, who doesn't see her daughter in a dual-language program, but who would like to find and utilize second language learning resources, classes, and teachers to support what they do at the suburban schools.

CHOOSING AND USING PROGRAMS: KEY TERMS AND DEFINITIONS

The number of terms out there describing different language learning programs for school-age children can be pretty daunting. These terms are not always used consistently, and to make matters worse, there is a lot of overlap between different types of programs. So, before we begin to look at the pros and cons of various types of programs, an overview of some basic terminology is in order.

What is bilingual education? Actually, there are lots of different types of bilingual education programs. Some of the most well-known ones are: maintenance, transitional, and immersion programs.

It's important to keep in mind that while the goals of most of these programs are to teach children more than one language, they vary considerably in how they are implemented across schools, school districts, and states (or provinces). Some of the differences across and within programs include:

- When the second language is introduced (for example, in kindergarten or later on)

- Whether the child's first language is used, and if so, how much and for how long
- Whether the second language is used by a minority or a majority of the surrounding population
- What the goal is (bilingualism or a transition to one language only)
- Whether there are legal restrictions on how bilingual education can be implemented (more on this in a minute)
- Who is eligible and how students can enroll

Maintenance Bilingual Education Programs
Goal: Mastery of Both English and Another Language

In maintenance bilingual education programs, children receive instruction through the medium of both languages, typically beginning with instruction in the native language only (again, usually Spanish in the United States). Instead of phasing out the first language (as transitional programs do), the second language (English) is *added*. The exact time when English is added (for example, first grade, second grade, and so on) and in what proportions vary from school to school. The overarching goal of a maintenance bilingual education program is not just mastery of English, but mastery of English *and* another language. These maintenance programs have been found to be effective in maintaining students' native language and also in promoting academic achievement *in English*.

Both maintenance and transitional bilingual education programs have come under attack in recent years. For instance, both California (Proposition 227) and Arizona (Proposition 203) voters have passed referenda that have sharply limited use of languages other than English in public schools. Partly for this reason, bilingual education programs are relatively rare in some places. For instance, in California, before the anti-bilingual education Proposition 227 was passed, about 30 percent of English language learners were enrolled in bilingual programs. In the following year, only 12 percent were.

Fortunately, it's not like this everywhere. There are other states where bilingual education is on the rise, and programs can still be found in Arizona and California. Still, depending on where you live, it can be difficult to find a public school system that has a suitable bilingual education program. Later in the chapter, we provide lists of resources to help parents find different kinds of bilingual programs for their children. Let's turn now to transitional programs, which are far more common.

Transitional Bilingual Education Programs
Goal: Transition to English

In the United States, transitional bilingual education programs are by far the most common form of bilingual education. Here, students receive content matter instruction in their native language (for example Spanish, Chinese) for a relatively short period of time (anywhere from a few months to a few years) with the aim of helping them keep up with their English-speaking peers in subject-matter such as math and science. Students also receive special English as a second language (ESL) instruction. Later on, typically by the end of the first or second grade, they are transitioned to mainstream (that is, English-only) classes. They may continue to receive additional ESL support classes. The goal of this kind of program is fluency and literacy in English, and not bilingualism.

Why is this form of bilingual education so common? In 1974, the U.S. Supreme Court ruled in the case *Lau v. Nichols* that some schools in San Francisco, California, were violating (approximately 1,800) Chinese-speaking students' rights by not making provisions for the fact that they didn't speak or understand English. Simply providing students with the same curriculum, textbooks, teachers, and facilities as monolingual English speakers was judged to be

· **FAST FACTS** ·

TYPES OF BILINGUAL EDUCATION

Maintenance bilingual education: These programs are to help children become both *bilingual* and *biliterate*. In the United States, students in such programs typically speak a language other than English at home.

Transitional bilingual education: These programs aim to help children transition from their native language (for example, Spanish, Cambodian, Portuguese, Arabic) to the language of the majority culture (in the United States, English). Content matter is taught in the child's first language initially and the child simultaneously receives instruction in English as a second language. Later, the child is moved into classes taught in English for all subjects.

Immersion bilingual education: Students are generally native speakers of a majority language (for example, in the United States, this means children are English speakers), and 50 percent or more of the content matter is taught in a second language (percentages vary across schools). The idea is that students are fully immersed in the second language throughout the school day.

Two-way (or dual-language) immersion bilingual education: These programs aim to help native speakers of a language other than English (such as Spanish) to learn English, while at the same time helping children who already speak English to learn this other language. Children from both language groups are together much of the day, and content matter instruction is delivered in both languages. The goal of these programs is to help promote bilingualism, biliteracy, and cross-cultural understanding for all.

Source: http://www.ncela.gwu.edu/expert/glossary.html

insufficient if the students did not understand the language of instruction. In other words, so-called sink or swim approaches were effectively preventing some students from receiving a meaningful education and so unacceptable. The Supreme Court ruled that the school district had to take measures to identify non-English-speaking students, to provide appropriate educational services for those students, and to teach them English where necessary. Exactly how school districts did this was up to them (which is one of the reasons that programs have taken such vastly different shapes across the country!).

Transitional programs, like all bilingual education programs, have been at the center of much controversy ever since. Some have claimed that children in these programs are cradled for too long in their native language and that their learning of English is thereby hampered. These people often argue that children who do not speak English as their native tongue should be placed in English-only, mainstream classes from the very beginning of their schooling (i.e., no bilingual education, period). Others note that transitional programs' focus on English-language prevents children from becoming literate in their native language and don't provide adequate support for the development of academic English skills. They suggest that while the development of basic oral communication skills (sometimes known as playground English) only takes a few years, learning the *academic* forms of a particular language (the concepts and skills needed for successful academic communication in the classrooms) takes much longer—between five to seven years. What's more, research suggests that transitional programs are less effective in the long run than maintenance programs, which are more effective in promoting language skills and in teaching academic content.

SPOTLIGHT ON RESEARCH:

Transitional Bilingual Education Programs

In 2005, Kellie Rolstad and colleagues conducted a review of research studies looking at the effectiveness of various types of programs for second language learners. Looking at seventeen different studies, these researchers found that for English language learners, transitional bilingual education programs were actually among the least effective in terms of promoting academic achievement. Programs that sought to help children become bilingual and biliterate (that is, maintenance programs) were more effective than both transitional programs and English-only programs in overall academic achievement. Among the conclusions drawn from the research review was that the recent move of states such as California and Arizona to severely restrict bilingual education was ill-advised. The authors recommend that more maintenance bilingual programs be developed in the future.

While the debate is far from being resolved among educators, parents who speak a minority language in the United States may wish to explore alternatives to the transitional model if they are hoping to raise *bilingual* and *biliterate* children. Transitional programs, at their heart, are geared toward monolingualism and English mastery. The maintenance and development of another language are not part of the plan. In addition, since most (but certainly not all) of the transitional programs in the United States are geared toward native Spanish-speaking children learning English, they may not be a good fit for families whose home languages are other than Spanish and English.

> • FAST FACT •
>
> Transitional bilingual education may not be the best fit for parents who want their children to know two languages well.

Immersion Bilingual Education Programs
Goal: Bilingualism, Biliteracy, and
Mastery of Students' Non-Native Language

Immersion is another major type of bilingual education. There are different flavors of immersion, with labels like *total immersion, partial immersion,* and *two-way immersion.*

- In *total immersion,* subject-matter instruction is provided completely in the foreign language—especially in the lower grades. Later on, English instruction is generally introduced and (depending on the program) may increase to make up about 20 to 50 percent of instruction, creating more of a balance between English and the other language.
- In *partial immersion,* instruction is provided partially in the foreign language (for example, for up to half of the school subjects) right from the start. In some programs, teachers reinforce English concepts that have been taught in the foreign language.
- *Two-way immersion programs* (also known as dual-language immersion) take a slightly different approach in that they are designed to help native speakers of English become fluent in another language (for example, Spanish) and native speakers of non-English languages become fluent in English. To this end, there is a balance in the number of minority and majority language students who are integrated for much of the school day, and all students receive content and literacy instruction in both languages.

The goal of all three types of immersion programs is to promote high-levels of bilingualism, biliteracy, and cross-cultural understanding, and there is evidence that they are successful at doing all of these things. For example, as we mentioned in chapter 1, there is evidence that children who participate in dual-language immersion programs have more positive attitudes toward speakers of other languages (relative to children who don't participate in such programs), which in turn helps them to make friends of different ethnic and linguistic backgrounds. In addition, there is also a lot of evidence that these programs help children become both bilingual and biliterate, *with no cost* to their academic skills and content knowledge.

 SPOTLIGHT ON RESEARCH:

Two-Way Immersion Education Programs

In a recent (2002) study, Barbara Senesac looked at one highly successful two-way (Spanish-English) immersion program: the Inter-American Magnet School in Chicago. Over a ten-year period, Senesac collected data from a variety of sources, including observations of classes, staff meetings, committee meetings, and parental advisory meetings. She also conducted extensive interviews with teachers, administrators, parents, and students, and looked at students' scores on standardized tests and their English and Spanish proficiency. She found that the students in the immersion school scored at or above district and state averages, and that they had good proficiency in both languages. Discussing what made the Inter-American Magnet School so successful, Senesac pointed out the following characteristics: "(a) a challenging core curriculum; (b) a nurturing, family atmosphere with high expectations for learning and personal development; (c) a dedicated, collegial, and highly trained staff; (d) pedagogical approaches and strategies that are student-centered, fostering interaction and active engagement in learning; (e) a thematic curriculum reflecting the culture of the students; and (f) parent and community collaboration."

HOW DO I FIND ONE OF THESE
GREAT IMMERSION PROGRAMS IN MY AREA?

Fortunately, there are databases of foreign-language and two-way bilingual immersion programs in the United States, which can be found at http://www.cal.org/resources/immersion/ and http://www.cal.org/twi/directory/, respectively. Here, you can search by state or by language to find an appropriate immersion program. The organization that runs this Web site, the Center for Applied Linguistics (CAL), also provides other pertinent information, such as the number of two-way immersion programs in different languages. As of late 2006, there were about 250 foreign-language and 330 two-way immersion programs represented, and these numbers are growing fast due to parental demand.

In addition to CAL, there are a lot of other Internet resources that can be found through search engines like Google to look for specific languages. For example, typing in "Chinese language program" brings up a helpful list of Chinese language programs in public and private schools in the United States (http://www.internationaled.org/uschina/K12Chinese Programs.xls). The Web sites of local school districts can also provide a wealth of information on the resources available in their area. For example, the Web site of Fairfax County Public Schools in the state of Virginia (http://www.fcps.edu/DIS/OHSICS/forlang/partial.htm) presents lists of schools offering various kinds of immersion programs in the county (along with their contact information), descriptions of different program models and their goals, schedules of informational meetings, enrollment procedures, and application forms. You can check for similar kinds of information where you live.

You may be a bit concerned to find out how difficult it can sometimes be for parents to ensure that their children get spots in a particular two-way immersion program. And it's true that in many places, supply of such programs doesn't keep up with demand.

However, it is also the case that immersion programs in other places are actively recruiting for enrollment. Don't lose hope; just start researching early and be proactive. You might even consider re-reading the Which Language and When? chapter, and remember there are many options. The key is to be flexible!

We should also mention that although the number of immersion programs in languages other than Spanish and English is increasing, the vast majority of two-way immersion programs are Spanish-English. For parents who are interested in other languages, it may be more useful to explore alternatives to immersion programs.

IF BILINGUAL EDUCATION IS NOT GOING TO WORK OUT, WHAT ABOUT A DIFFERENT KIND OF INSTRUCTIONAL PROGRAM?

There are lots of other ways to provide children with exposure to a second language. These include self-contained foreign language classes within most public and private schools. There are also heritage language schools and summer camps, private language schools, and classes sponsored by community agencies, cultural associations and centers, and religious organizations. If you are mainly interested in building your child's cultural awareness through exposure to another language, you might be very satisfied with the foreign language classes that are sometimes offered as part of a school's regular curriculum. In these classes, which often meet for a little less than an hour a week, young children learn colors, numbers, songs, and common expressions that they can use with peers who speak other languages. This sort of exposure can create positive associations with the language and possibly build a bit of an initial foundation for later, more rigorous academic study. However, if you are more interested in your child developing the ability to use a language fluently for flexible and effective communication early on in life, you need to look for more than this sort of introductory class.

Heritage Language Classes, Schools, and Teachers

Heritage language schools (or classes) provide some options along these lines. These programs are created by community organizations to help children learn the community's ancestral or heritage language. There are an increasing number of heritage language schools in the United States and elsewhere, with languages such as Chinese, Japanese, Korean, Vietnamese, Persian, Russian, Navajo, and Cree (among others) being frequently targeted languages. (More details on how to find them in our resources section at the back of the book and on our Web site.) Some of these types of programs have been around for as long as their associated ethnic or cultural group has been living in the United States, and at least some of them have historically received funding from their countries of origin. Classes are often offered after regular school hours or on the weekend to help children learn another language after their school day is completed.

Since these are often nonprofit organizations with a limited budget, parents often serve as board members, as part of the school administration, and even as teachers, in volunteer positions or with minimal compensation. In a recent survey of Chinese-language school teachers, over 50 percent of the teachers interviewed were also parents. Many schools have a fee for tuition. For example, in Minnesota, the Rochester Chinese School and the Twin Cities Chinese Language School in Saint Paul charge approximately $100 per semester. Other programs may cost more or less depending on the location, size of school, whether the teachers are paid for their time, etc.

It is important to note that most heritage language programs are designed to assist children who already have at least some exposure to the target language (or a related dialect) at home. Heritage programs provide a different kind of instruction compared to language programs that are designed to support the learning of a foreign language and culture from scratch, so to speak. Using the knowledge children already have as a foundation, most heritage language programs try to boost or maintain children's communication skills, develop literacy (in school-age children), and strengthen cultural awareness. In addition to focusing on language proficiency, they provide a place where children can share their ethnic values and explore and reinforce connections with different facets of their cultures and identities.

Many Chinese heritage language schools have evolved over time to adapt to the needs of the community. While these schools used to only serve the children of Chinese and Taiwanese immigrants or those of Chinese or Taiwanese descent, recently, many of these schools have expanded their programs to include courses for adopted Chinese children and their parents as well as many non-heritage learners. Many of these schools also offer courses in subjects such as calligraphy, dance, art, or martial arts, mostly taught in Chinese.

In considering a heritage language school for your child it is important to keep in mind that since many of these programs are nonprofit organizations run mainly by volunteers, the quality of the instruction can vary. The teachers may not have formal training in teaching the target language—they could be native speakers who are parents themselves, or members of the community. However, they may still be enthusiastic, committed teachers. Visiting the school and interviewing the teachers will tell you about this.

One special category of heritage language programs is the kind of program that exists primarily to serve the children of foreign nationals who are likely to return to their country one day. They are designed for parents who want their children to follow the curriculum of that country, as well as to have language instruction in their native language. For example, the British schools of America organization operates private K–12 schools in Charlotte, Boston, Chicago, Houston, and Washington, D.C. The school operates in English and mirrors the British curriculum.

Some heritage schools also enroll students whose parents will not be returning to any other country. For example, the Washington Japanese Language School provides opportunities for children to learn Japanese as their heritage language. The program's main aims are to maintain and develop children's Japanese language skills and to teach children cultural values and manners. Programs are available for students from preschool through high school, and classes are held every Saturday from 9:30 a.m. until 4:00 p.m. at a public school located in a suburb of D.C. This school receives some of its funding from the Japanese government. The students learn Japanese, math, and other subjects (such as science and history) and attempt to keep in step with how these subjects are taught to Japanese students in Japan. All of the classes are taught in Japanese by a native speaker of Japanese with the same textbooks used in schools in Japan. As well as these classes, the school also offers cultural events to the students (for example, sports days), similar to those offered by typical Japanese schools. The school emphasizes that in order for the students to become fluent speakers, their parents need to be actively involved in promoting Japanese learning outside of school. With this in mind, it offers special events for parents and tries to serve as a place where parents can expand their resources and connections with people who share similar cultural values.

Summer Language Camps for Heritage Language Programs

Besides the heritage language programs that can be found in school and community settings, there are also "heritage camps." Many of these are geared to families who have adopted children from countries in Latin America and Asia (in particular, Cambodia, China, India, the Philippines, Russia, Korea, and Vietnam). These camps provide opportunities for children and their parents to learn not only the language, but also dances, folktales, traditional songs, and arts and crafts. Sometimes they also provide workshops to address self-esteem issues. In a typical Chinese heritage camp, for example, cultural experiences are provided for all family members. Families with young children adopted from China can participate in classes and workshops taught primarily by Chinese Americans for their children, where they can learn Chinese dances, traditional playground games, martial arts, and Tai Chi, in addition to learning about traditional holiday celebrations.

Although a camp in summer by itself is not enough to develop high levels of second language knowledge, it can be an excellent support for other sorts of yearlong programs, and for some families, what a language camp provides can be just what they are looking for.

Cultural Associations and Clubs

Another way for children to develop skills and knowledge that can be helpful in maintaining connections to their cultures and identities is provided by cultural associations, clubs, and centers in local communities. These organizations offer a variety of cultural events and classes, exposing children to their heritage culture through activities related to music, arts, food, sports, and religion. Cultural connections can enhance language learning and be very motivating for learners. While we parents understand the benefits of knowing a second language, we need to remember that for our five-year-old, being able to prepare a traditional dish, or having a regular Saturday dance class with bright costumes, music, and parent participation, can be the motivator needed to keep them going.

If you look, there are a lot of cultural associations and clubs. For example, the Arab Cultural and Community Center (ACCC) in the San Francisco Bay area was established to promote Arab culture and provide resources for the local Arab American community. It offers weekly programs such as guest speakers, films, and musical performances. It also hosts an annual Arab Cultural

Festival showcasing arts, traditions, and forms of entertainment of the Arab world. In many cities, the Alliance Française offers cultural events for toddlers to teens, including story time, arts exhibits, movie clubs, and all sorts of fun activities.

 QUICK TIP: Check Out the Web!

There are lots of lists of Web sites of cultural associations, clubs, and centers across the United States. Many useful links can be found on our Web site (www.thebilingualedge.com).

Private Language Schools

In addition to heritage language schools, camps, and other heritage programs, there are also private language schools that offer classes in various languages for both children and adults. Although they are typically more expensive than other options, private language schools can also be a valuable resource for busy parents.

In the United States there are no centralized associations or accreditation processes for private language schools although in Canada, the Canadian Association of Private Language Schools provides an exhaustive list of its member programs that teach English and French. What this means for parents is that you should first seek out schools that are convenient and local, look at their programs and schedules, and then ask for further information following our guidelines below, to find good private programs.

QUICK TIPS: Finding Schools and Programs

• Contact your city council for a list of language schools and classes in your area.
• Contact national or community organizations directly for information on schools and programs.
• Contact embassies or consulates for information on schools and programs.
• Contact the foreign language departments of local colleges and universities—often faculty there will have information on local resources for various languages.
• Let your fingers do the walking: Look under "Language Schools" in the Yellow Pages.
• Use the Internet (you can often search for programs using interactive maps and filtering programs according to pre-specified criteria).

WHAT DOES A GOOD PROGRAM LOOK LIKE?

Well, we started out this chapter saying that parents may wonder, when navigating this sea of information on different schools and programs, what a good program looks like. Good programs come in many different shapes and sizes, and what is good for one family's set of needs may not be good for another's. Having said this, though, asking some questions and doing a little legwork can help you to determine whether a program is worth your investment of time and money.

One of the first and perhaps most important questions parents should ask concerns the philosophy of the school or language program. Good programs usually specify what level of proficiency or bilingualism they expect, and many also have biliteracy as one of their main goals. If the school does not seem to have clear objectives or goals, you may wish to look elsewhere. Moreover, good schools respect cultural diversity and seek to promote cross-cultural understanding—not just (as was sometimes the case for bilingual programs in the past) assimilation to the larger majority culture. Many of us would like to believe that all programs these days are actively promoting bilingualism, diversity, and the recognition of different cultures, but the recent debates concerning bilingual education testify to how politically charged such issues can be.

In addition to asking administrators and teachers about these issues, visiting the school itself is invaluable. Ask if you can observe a class or have a tour of the school. While you are observing, take notes on everything you see and hear, paying particular attention to the issues below.

Curriculum and Lesson Plans:

Asking to see the curriculum and specific lesson plans can also give you an idea of the amount of time spent on the foreign language and the overarching goals of the program. Although it seems obvious to say this, if a school offers mostly lessons on popular topics like colors, numbers, and days of the week in the foreign language, and the vast majority of instruction is in English only, you will probably wish to look elsewhere. While a few weekly lessons of another language may be a good start, if your goal is for your child to be very fluent or actively bilingual, then the amount of language exposure and communicative interaction is very important. You didn't pick up this book expecting that half an hour of French a week was going to make your child able to understand *Madeline* in French by age six, so neither should you think any school or program can work that kind of magic without more language input than a lesson a week.

Materials and Environment:

What sort of materials are being used in the lessons? Do they use a variety of things designed to appeal to visual and auditory learners (for example, snippets of DVDs and other input, as well as the usual arts, crafts, and books). Is some or all of the play time conducted (or mediated) in the second language? Are there songs, stories, and games in the second language? Also consider how the classroom is set up: Are the chairs and tables movable (for small group work)? What sort of posters and pictures are on the walls, not only in the classroom but all around the school? All these things are beneficial and should be contrasted with a formal, "drill and kill" approach, which can be boring (if not fatal!).

Native-Speaking Teachers:

You might also consider how many native speakers of the target language the school employs, and ask those teachers a few questions if possible. You will want to get an idea of the school or program's atmosphere. As we have said before, you don't need to be a native speaker to provide quality second language input and interaction, but in a language program, having a few of them, or at least near native speakers, can be a good sign.

Teacher Qualifications, Experience, and Retention:

What sort of qualifications do the teachers have? Do they have a background in language teaching? Child education? How long have they been teaching this age group? What is the school's turnover rate for teachers? Do teachers get down to the level of the children and play and talk directly with them using the second language?

Fun with the Language:

Are the teachers and students *actively* using the language (and not just listening to lectures, working quietly at their desks)? Are they motivated and enthusiastic? Are there plenty of different kinds of game materials in both languages?

Extracurricular Activities:

Does the program offer any sort of extracurricular activities or support? In a school or program that is very committed to its goals, strong participation in extracurricular activities can be a sign of interested and involved parents.

SPOTLIGHT ON RESEARCH:

The Value of Actively Using the Language

In one early and important study, Merrill Swain (1985) examined native English-speaking students who were studying French in francophone Canada. The learners were exposed to French daily both in their classes and in the surrounding community, but the curriculum did not place much emphasis on active speaking and writing in the language, and the students' comprehension abilities far outstripped their speaking and writing abilities. While they could understand everything that was said to them in French at a nativelike level, they still made a variety of mistakes when they talked. A recent graduate of this immersion high school described her experience by saying, "You go off and read, and that was about it. We read. That's all I can remember, is reading." She also pointed out that when she spoke French in class, she would sometimes sneak in words in English here and there, but her teachers never really called her on it or pushed her to use French exclusively. In light of these findings and many similar anecdotes, Swain argued that it is not enough for a program just to expose students to a second language. Rather, students need to be given opportunities to produce the language actively. They need to practice using their linguistic resources in meaningful ways to become fluent. They may also sometimes need to be forced to pay attention to more than just the meaning of what they are hearing, focusing specifically on grammar. Whereas it's often possible to use contextual clues and world knowledge to fill in gaps while listening, the act of trying to produce language can help make learners aware of what they don't know. This is something to think about if you're trying to evaluate an immersion program for your child. Take it from the Canadians—they've been studying how immersion works (and how it sometimes doesn't work) for decades!

• FAST FACTS •

GOOD PROGRAMS HAVE:

- Clear goals that include bilingualism, biliteracy, and biculturalism
- Motivated and enthusiastic teachers and students
- Students and teachers who actively use the target language
- Support of their communities
- An open door policy and will not make you feel like a pest for asking questions.

Other questions to ask concern the school or program's sources of support. While some successful programs operate on a shoestring budget, asking about the financial backing of the program will give you another piece of information that may help you make a decision.

You can also ask about community support for the program. Visiting community centers, churches, or even parents who have children at the school can help you to determine whether the school has a good reputation, resources, and the backing of the community. Increasingly, online message boards where parents communicate with each other can provide similar sorts of information. It's important to keep in mind that even for some programs that do not have a strong financial base, this can be more than made up for by parental and community commitment and involvement.

INSTEAD OF (OR AS WELL AS) A PROGRAM: FINDING INDEPENDENT LANGUAGE TEACHERS AND TUTORS

For logistical or other reasons, parents are not always able to find a good program in their area. In such cases, they may wish to consider finding a language teacher, or tutor, for their child—someone who can provide one-on-one support in their child's language learning efforts. Basically, this means some individual instruction, as opposed to an organized program. A midway option: Families often band together to get a tutor for a small group of their children, holding classes in one of their houses.

As we all know from our own experiences as students, there are many different kinds of good teachers. Your favorite teacher may have been someone who was soft-spoken but kind, while another's favorite teacher was gregarious and had a great sense of humor. However, while there is no set list of criteria to identify a good language teacher, it can be extremely helpful and enlightening for parents to sit down and talk with teachers in person (rather than on the phone). This can help you find out what kinds of teachers (and people!) they are and whether their views match your own with respect to languages and learning. We mentioned in the myths section of this book that it is not necessary for children to interact only with people who always speak in complete sentences (all sorts of input and interaction are useful), nor is it essential that a child's teacher be a native speaker. However, there are some characteristics that may be particularly helpful. Some sample questions you could ask include:

- What is your philosophy or general approach to teaching language?
- What benefits do you think children gain from learning another language? Do you see any drawbacks?
- Is there anything special about teaching children (or adolescents)?
- What teaching credentials do you have?
- What experience do you have teaching this language and age group?
- What's your opinion on explicitly teaching grammar?
- Are there any ways in which you try to stay up-to-date with current ideas about effective teaching and how children learn languages?
- What do you do in a typical lesson or session?
- What sorts of activities and materials do you use?

- How do you try to ensure that your students get practice actively using the language?
- What are some of the things you do to encourage and motivate your students?
- What role do you see for culture in language classes?
- How do you deal with students' errors?
- How does your approach change depending on students' individual differences?
- What suggestions do you have for my child to keep up the language outside of your classes?
- Can we see a sample lesson or a video of you teaching so we can get a better idea of how you teach?
- What kinds of materials do you usually use in your lessons?
- What do you do to engage the learner in reading and writing? In speaking and listening?

In general, teachers who are enthusiastic and committed to the task of helping children become speakers of more than one language should go to the top of your list. Teachers who indicate that their lessons are not just lectures and grammar drills also deserve a top position because, as we discussed before, learning a language involves *actively using* that language and not just listening to it or working through problems about it. In addition, teachers who have experience studying a second language and who have credentials (for example, in the form of a teaching certificate or even a recommendation letter from the parents of previous students) are always worth consideration. Perhaps above all, a good attitude is of paramount importance when looking for a teacher. Teachers with positive attitudes (in other words, teachers who believe that every child can learn and that every child has the potential to succeed) are the ones who will go the greatest distance to help your child.

QUICK TIPS: Sifting Through Candidates

Does the teacher or tutor you're considering have the following?
• A positive attitude toward language learning and learning in general
• Recommendation letters from previous students or parents
• A plan for each lesson they conduct, as well as backup activities
• A philosophy of teaching that places value on active participation of the learner (rather than just lecturing about grammar points, for example)
• Experience working as a tutor or teacher for children or adolescents
• Experience learning a second language him- or herself
• An affinity for communicating with children in same age group as your child
• Some recent coursework or professional development activities in relation to second language education

Teachers' Portfolios and Observations

Many prospective teachers will present portfolios when you interview them. These portfolios often include resumes, teaching plans, recommendation letters, certificates, videotapes, and other evidence of how they teach. Portfolios are becoming standard parts of the application process for many teaching positions, and an experienced teacher or tutor is quite likely to be familiar with this system. Sometimes, teachers have also collected portfolios with copies of their students' work, which can give you a sense of the sorts of projects they assign and how enthusiastically and skillfully their students complete them. Looking through their portfolios with them can be very helpful, especially listening to them talk about various pieces. In some cases, portfolios are exit requirements for teaching qualification programs and are more boilerplate than enthusiasm. Ask teachers about their plans, why they choose the topics and activities they did, how and why they developed their portfolio and how their portfolio reflects their teaching today.

You may also want to ask to observe the first few lessons that a teacher gives. This can give you a sense of whether there might be a good match between the teacher and your child, and can also help you get to know the teacher better. Don't be shy about asking to observe lessons—if someone is going to be spending a good deal of time with your child, it is only natural that you might want to see him or her in action. Most qualified teachers will be willing to allow you to observe. A word of caution though: When you observe a lesson, you are a guest. Some teachers and tutors will not appreciate being interrupted or having you put in your two cents' worth during the lesson. It would be good to save any comments or questions until after the lesson, when you and the teacher have a bit more time and you won't interrupt the flow of their lesson. You also want to model good behavior and not take away focus on the class for the children involved, so being polite and unobtrusive is usually appreciated by teachers who are being observed.

When parents try to find language learning opportunities for their children outside of the public school system, there are a number of things to think about regarding tutors. Obviously, finances also need to be considered. And don't forget that an excellent way to deal with this issue is for friends and neighbors to gather together and make up a small class, which then becomes more affordable. In chapter 6, we also talked about the practicality of hiring nannies and babysitters who speak a second language.

The Methods Don't Fit! What Do I Do?

Bob and Melinda recently hired a tutor for their eight-year-old son, Andrew, in order to help him learn French. After observing the teacher for a few lessons and hearing their son's comments about the teacher (for example, "He's boring—he just talks about verbs all the time."), they are worried that the tutor may not have been the best choice. They don't want to offend the tutor, though, and they are not sure how to broach the issue diplomatically. What should they do? Bob and Melinda are facing a situation that many of us regularly encounter in one form or another. We buy something and it doesn't fit. We hire someone, and they don't do as well, or exactly what, we expected. We make a decision and later realize it may not have been the best choice. The same thing can happen when hiring a tutor or teacher. Alison, for example, continued to employ a monolingual part-time nanny who was unreliable, even after this nanny had obviously become uninterested in much of anything but collecting her cash, because Alison's daughter Miranda seemed to really like her. However, after the nanny finally departed, Miranda never mentioned her again, her language skills picked up with different babysitters, and the whole family was a lot happier. In hindsight, Alison wishes she had unstuck herself and her daughter from the nanny sooner.

One strategy when faced with a tutor who seems conscientious but whose methods just don't seem to fit with your ideas about what's best for your child would be to sit down with the tutor and explain that you were expecting a slightly different approach to language learning. For example, Bob and Melinda could mention to their child's tutor that, while they have faith in his expertise, they aren't sure that the best way for Andrew to learn is by spending an entire hour listening and practicing sentences in the past tense. They could also ask to discuss the tutor's longer-term plans. After all, he could simply have been trying to lay a foundation in that boring lesson, and might have a lot of fun activities for the future up his sleeve. If that does not appear to be the case, Bob and Melinda could ask the tutor to reconsider his teaching approach. If he seems unwilling or unable to comply, then, of course, they have the option not to continue and to part amicably, recognizing that they simply have different goals or expectations in mind. In our experience, though, many tutors are ready and willing to accommodate parents.

If you find yourself in a similar situation, perhaps you could give the teacher this book (highlighting certain sections) to show him or her where you're coming from. In the process, you might not only help to shape the tutor's teaching style, but also create a more open and cooperative relationship in general. It's important to bear in mind though, that just like our children, tutors come in many shapes and sizes. Tutors from some language backgrounds might be a bit more authoritarian than others, because that's how they were taught, for example. But if you really feel as though this is not a good match for your child, finding an alternative is probably best.

In practice, we do realize this can be difficult to do. (Alison, after all, never got around to letting that not-so-great babysitter go; she finally left of her own accord). And we don't want to leave you feeling overly optimistic, either: Some programs, tutoring organizations, and even some tutors ask to be paid up front and will not give refunds. This is why you should carefully investigate the teacher, methods, and modus operandi in general before beginning. In our final chapter, where we talk about problems and resolutions, we discuss these situations in more detail.

WRAP UP

An incredible number of resources exist outside the family to help parents in their efforts to raise multilingual children, including programs in public school systems, heritage language schools and camps, and classes sponsored by cultural associations, community agencies, and religious organizations. In addition, private teachers are available to help your child. In sifting through the available resources, it is important to determine which ones are truly committed to the development and maintenance of a second language. Fortunately for parents, the number of institutions that are truly geared toward helping children become bilingual, biliterate, and bicultural is on the rise. And if such programs are not currently available in your area, you can always try to mobilize other like-minded parents to create resources in the community by actively petitioning for their development.

One place to start might be with Marcia Rosenbusch's "Guidelines for Starting an Elementary School Foreign Language Program" (available at http://www.cal.org/resources/digest/rosenb01.html). As Rosenbusch points out, it is important for planners to be aware of factors that have tended to stand in the way of success in the past, including such seemingly basic things as whether teachers with sufficient language skills and qualifications can be found, how realistic the program goals are, and how the program can be evaluated and coordinated across levels of instruction. A steering committee should be formed to lead the process, and it should be made up of representatives from all stakeholder groups, including parents, teachers, administrators, and business and community members. It will be their task to research and convey the rationale behind the program, to weigh the pros and cons of different program models, to explore available support, to assess feasibility with respect to the local context, and to get the community informed and involved, among many other things. Determining the languages of instruction may turn out to be one of the most controversial issues, so interested parties might want to be ready for some pretty serious negotiations. The Web site provides a list of useful resources to help you get started. Undertaking something like this isn't exactly a piece of cake, but the outcome can be extremely rewarding and even inspiring if you've got the will and the wherewithal to take it on!

EXERCISE:

Brainstorming Questions to Ask
Schools and Teachers

In this chapter, we have listed some questions that may help you to identify good programs and teachers. However, each family will have their own questions. Take a minute to brainstorm your questions here, also keeping in mind practical considerations like schedules (yours and theirs), costs, and transportation, as well as the intangibles that contribute to the overall environment of a school, such as collegiality among the faculty and staff, demographic characteristics of the students and teachers, high expectations for learning and personal development, a nurturing and student-centered atmosphere, cultural awareness, and general levels of energy and enthusiasm.

SECTION FOUR

What If . . . ?

CHAPTER 9

What if My Child Mixes and Switches Languages?

anguage mixing is often a big concern for many parents and one that causes no small amount of anxiety for other family members as well. We've even worked with parents who unnecessarily have postponed or altogether abandoned their bilingual plans because they find their child's mixing so disconcerting. Other parents are anxious that young children might become so overloaded with two languages that they might not be able to differentiate between them, or that they might not understand that two different words can refer to the same concept or idea. Others are sometimes concerned about their children's impromptu and apparently unconscious mixing of languages. What if it never gets sorted out?

Linguists and language researchers have tried to answer some of these concerns through systematic study of how and why children and adults mix languages. While there are still active, ongoing investigations in this area, research suggests the following five points very clearly:

- Don't worry if your child mixes languages—language mixing is a common (and typically short-lived) phase of bilingual development.
- Trust your child is not confused—she may not know (or be able to explain) that she's using two languages, but there's plenty of evidence to suggest that she has two linguistic systems.
- Understand a bit about how and why children mix when evaluating your child's language use.
- Minority languages may need extra support, and frequent use of both languages together can make it difficult to keep an eye on the support for each language.
- Set realistic expectations for your young learner—there are no perfect bilinguals in the world, and remember that language learning is a lifelong process—it's never done.

LANGUAGE MIXING IS A COMMON PHASE OF BILINGUAL DEVELOPMENT

Whether we are giving a professional talk, appearing on the radio, or having a chat at the playground, worries about mixing (along with worries about language delay) are among the most common parental concerns we hear. And, indeed, children who are learning two languages do mix them up at times. Anyone who is around young children learning two or more languages can attest that they occasionally combine words or structures from two languages together into one sentence. A short example of a two-year-old (Tania) with her Spanish-speaking grandmother. (Spanish in italics; translations in parentheses.)

Grandmother: *Vamos al parque.* (Let's go to the park.)
Tania: Park, park, park! *Sí. Sí. Sí.* (Yes. Yes. Yes.)
Grandmother: OK. *Vamos.* (Let's go.)

Tania: Yea! *Yo quiero ir al* park. (I want to go to the park.)
Grandmother: OK. *Vamos.* (Let's go.)
Tania: *Vamos a ver los* doggies. (We'll see the doggies.)

In this example, we see that two-year-old Tania moves back and forth between two languages. Her sentences here are mostly in Spanish:

Yo quiero ir al park.
Vamos a ver los doggies.

And she uses Spanish to express some of her enthusiasm about the park (*"Sí. Sí. Sí."*). However, Tania also expresses her happy anticipation in English ("Yea!"), and it may seem somewhat striking that English words ("park" and "doggies") appear in sentences that are otherwise well-formed according to the rules of Spanish. Why might Tania switch this way? There are lots of different possible explanations. For one thing, while Tania's family speaks to her in Spanish, she lives in a wider community that speaks mostly English, so it's possible that she hears the English words "park" and "doggies" more often than the Spanish words *parque* and *perritos* and prefers the English words for that reason. Another possibility is that "park" and "doggies" are simply a bit easier for her to say than *parque* and *perritos* since the English versions are a bit shorter and simpler in terms of the sounds she has to make. Maybe she recognizes *parque* when her grandmother says it, but isn't used to producing the Spanish word herself and knows she'll be understood just as well by her bilingual grandmother either way. Or maybe she can understand and say the Spanish versions, but associates both "park" and "doggies" with the English-speaking friends she usually sees at the park. While we can speculate until the sky turns green, it is impossible to say for certain.

While the precise functions of code-mixing are often unknown, the end result is very clear: *Children move beyond this phase.* While it is often unsettling to parents, the code-mixing phase is typically short-lived and finishes long before formal schooling begins. It is not problematic in any way in the long run. Many studies have shown that children are very sensitive to the unspoken rules about which language or languages should be spoken to whom and when, and naturally sort this out on their own.

What's more, most children do this *without* any explicit help or teaching, just as they learn, for instance, to use the plural "s" without instruction. And as children learn language (sounds, words, and the rules to put them together to get what they want: for example, "I need ice cream now!"), they also learn how languages are used socially in interactions (for example, how people take turns in conversation, what the raising or lowering of one's voice communicates, and that saying "Excuse me" before asking for ice cream may be more effective at getting an adult's attention and acquiescence). Through observation and participation in the conversations around them, children learn which languages should be used with whom, and how.

MIXING IS *NOT* AN INDICATION OF LANGUAGE CONFUSION—RATHER CHILDREN SEEM TO DISTINGUISH THEIR TWO LANGUAGES RIGHT FROM THE START

For a while, this type of language mixing exemplified here was considered—by both academic researchers and everyday parents—to be a sign that young bilingual children did not differentiate their two linguistic systems early on. It was taken to mean that children like Tania were unaware of the fact that they had two languages or were unable to separate them. Even though this work has largely been discredited as we'll see below, it filtered into popular knowledge, so we are still left with the hangover of these old findings in the form of outdated concerns that don't apply to most children.

One big problem with research in this area is: How can we really be sure of what words children know? When children are between one and two years of age, parents have *some* idea of which words their children know based on what words they say, of course. But research has shown that children at sixteen months can comprehend from two to five times as many words as they can say. And as children begin to acquire words very quickly, starting soon after their second birthday, the task of getting a firm handle on what they know becomes harder still. At that point, children are picking up about two hundred words per month—this is why this period is called the vocabulary spurt. We were amazed at the things our two-year-old children pointed to and labeled correctly—just within the month or two it took us to write the first draft of this chapter. Who knew (certainly not us!) that our children had in their brains the words *trapezoid* and *combine harvester*? And while it is hard for parents to know which words their children know, it's even harder for researchers who tend to spend far, far less time with their little subjects. It is nearly impossible to know all the words that children know, not to mention whether or not they knew the same words in both languages.

A few researchers have tried to get around this problem by examining lots of everyday language in a small number of families over a long period of time. For example, Elizabeth Lanza, an American linguist living in Oslo, Norway, studied in great detail the language development of two children from different Norwegian-American families residing in Norway. Each of the children had an American mother and a Norwegian father. In both families, the plan was for the mothers to speak English with the children and the fathers to speak Norwegian. Lanza recorded the children in their homes interacting with their mothers and fathers regularly for a seven-month period beginning when the children had just turned two. She then transcribed all of these recordings and conducted a detailed analysis of the children interacting in a range of situations with both parents.

In a nutshell, Lanza found that bilingual children as young as two years of age were able to use Norwegian and English, and were sensitive to which language should be used with whom and when (in this case, English with the mother and Norwegian with the father). They were also aware of when it was appropriate to mix the languages (for example, at dinner in one of the families).

Lanza's findings that toddlers generally know when to use what language confirm our experiences with many young children, including our own. For instance, by twenty months of age, Kendall's son knew to use *árbol* with his mother and the English equivalent (tree) with his father. Alison's daughter Miranda knew that when it was Japanese night in her house, she would have more success asking for dessert if she requested it by saying *manju* and not "cake" (manju is a sweet red bean–based dumpling that Alison buys from the local Asian market).

Our colleague Miriam reported a similar strategy was used by her bilingual two-year-old granddaughter, Anabela, one day when she wanted a lipstick from her mother's purse. (Spanish is in italics; translation in parentheses.)

Anabela:	Lipstick. Mine.
Anabela's mother:	No, it's not yours.
Anabela:	*Es mío. Es mío.* (It's mine. It's mine.)
Anabela's mother:	*No es tuyo. En español, tampoco.* (It's not yours in Spanish either.)

These sorts of findings also fit with what we know about how children process even smaller bits of language at even younger ages. Research suggests that similar systematicity exists in bilingual children's early pronunciation missteps as well. Regardless of whether a child is growing up bilingual or monolingual, children's vocal mechanisms are not always mature enough to accurately produce all of the sounds in their language(s) accurately. Yet when young bilinguals begin to talk, they seem to maintain distinctions between their two languages, that is, they tend to have two distinct sound systems, for instance, one for Spanish sounds and one for English. One study examined the language of a two-year-old Spanish-English bilingual child named Fernando, who was unable to make the sound *f*. Let's think about this for a moment. Producing an *f* sound involves placing the top teeth on the bottom lip and letting air escape with a sort of hissing noise. (Try to make this sound now—it sounds a bit like the air being let out of a tire.) This might not be so easy for a young child to do! Because he couldn't quite master the *f*, Fernando employed different strategies—or substitute sounds—depending on which language he was speaking. When he was speaking English, he used something like the consonant *p* (bringing his lips close together and then releasing them with a burst). When he was speaking in Spanish, he sometimes used the same *p* sound, but then switched to another substitute sound altogether that involved letting air make a hissing escape through his two lips instead of through his teeth and bottom lip. He never used this particular sound in English though. In short, even though he was still using what sounded like adorable baby language to most adult ears, he had already learned to differentiate between the sound systems of the two languages (although he could not render them perfectly himself yet).

The important point to take away here is that children seem to develop two language systems right from the outset. They aren't confused. In fact, carefully conducted research has shown that young children distinguish early on between their two languages, picking up on rules for language use very quickly—much earlier than had long been thought. Mixing language codes is very common and is apparent even at the babbling stage, that is, long before children can say a word. When children are at the point of saying words and phrases, they are also learning when it is okay to use which language with whom and when it is okay to alternate back and forth and use the two together. In fact, this form of linguistic practice or mental exercise may even be linked to greater scores on certain intelligence measures (see chapter 2 for more details on that one).

CHILDREN'S (SHORT-LIVED) *CODE-MIXING* IS DISTINCT FROM THE *CODE-SWITCHING* OF PROFICIENT, BILINGUAL ADULTS

Code-switching is a specific type of code-mixing that is commonly used by adult bilinguals in certain situations and for particular purposes. Children learn about the rules for using language from observing and gradually participating in interactions at home, but they also pick up the rules for using languages—and how they are combined and used together—from their wider community. In other words, children are sensitive to how, for instance, their cousins and playmates at the park move back and forth between two languages, as well as to how their parents, aunts, uncles, babysitters, and grandparents weave their different languages together or keep them separate. In some communities where two languages are widely spoken, it is common to *code-switch*, that is, to move back and forth between two languages in a rule-governed way within the same conversational turn. Code-switching is what proficient bilinguals do to express themselves and complex ideas. In contrast, *code-mixing* is what learners do when they are acquiring two languages. As we'll see, this is an important distinction for parents to keep in mind.

Code-switching is very common in many bilingual communities and often serves as an important communicative and strategic resource for expressing various sorts of meanings. Ana Celia Zentella has described, for instance, how Puerto Ricans in New York City employ English and Spanish simultaneously in order to establish their identities, communicate effectively with different members of the community (for example, not only people who are long-time residents of New York, but also those who have recently moved there from Puerto Rico), and manage their conversations according to community norms as well as their own individual styles. Code-switching can be used to promote solidarity, to make a particular point, to get across subtle nuances that might not be possible to convey through one language alone, or to do many other creative and functional things with language. The example below is from Zentella's work on a Puerto Rican block in New York City. Here, Lolita is in a small store and has put 25 cents on the counter for a bag of chips she wants, but is unaware that the price has gone up.

Lolita: *Bolo, cobra esto.* (Bolo, ring this up.)
Bolo: *Esto vale treinta centavos.* (That costs 30 cents.)
Lolita: *¿Esto?* (This?)
Bolo: *Sí, treinta centavos.* (Yes, 30 cents.)
Lolita: *¿¡Esto?!* (This?!)
Bolo: Yeah.
Lolita: Are you sure?!
Bolo: Yes, I'm sure.
Lolita: Are you sure?
Bolo: Yeah, 30 cents. You don't have 5 cents, OK, here, you owe me 5 cents.

The shopkeeper in this excerpt, Bodeguero (or "Bolo"), was an older man who preferred to speak Spanish. Lolita, on the other hand, was an eight-year-old girl from the neighborhood. Zentella explains the language use in this episode as something like "follow the leader." Lolita is more comfortable in English than in Spanish, but she speaks Spanish with the adult out of respect for what Zentella describes as a community practice to speak Spanish on the street with people who are not well known. When the shopkeeper switches to English (to accommodate the child and explain the change in price of the item she was buying), Lolita switches to English, too, but not before the adult makes the switch. In this way, code-switching can be seen to meet complex social goals in the community.

The distinction between this sort of code-switching and code-mixing is important for parents to keep in mind. Many children go through a phase of mixing two languages—they'll grow out of this as they master both. Code-mixing results from lack of mastery. For instance, many of us who studied a foreign language in high school have struggled when trying to communicate using only that language with a more proficient speaker. Because of our incomplete knowledge, we have sometimes ended up saying things like, "*Quiero una habitación* non-smoking" (I'd like a non-smoking room), using the English word for "non-smoking" because we don't know or can't come up with the Spanish word for the same concept.

QUICK TIP: How to Interpret Mixes or Switches

If children live in a home or community where code-switching is common practice, they will likely code-switch as well. This is not a sign of incomplete mastery, but a rich linguistic and cultural resource. If children live in a home or community where languages are kept separate, their mixing is probably a short-term phase as they sort out the rules for when to use what and gradually become fully proficient in both languages.

The distinction between the two is important to understand because of all the anxiety about mixes and negative attitudes toward switching. We've even worked with bilingual parents who have been reluctant to introduce two languages right from the start because of the fear that their children will become mixers (or switchers).

In contrast to mixers, code-switchers employ their languages systematically to convey complex ideas about the world, as well as about themselves as bilingual individuals. We know that code-switchers are fluent speakers of both languages because their code-switching tends to follow complex grammatical rules. By rules, we *don't* mean grammar rules that we learned in elementary school, such as "use *whom* instead of *who* for an object of a preposition." Instead we mean systematic rules of use. There are no grammar guidebooks for how to code-switch, of course. However, by looking closely at how speakers use languages together, researchers have been able to identify some rules of use.

SPOTLIGHT ON RESEARCH:

What Are the Rules of Code-Switching?

To begin to answer this question, in 1980 Shana Poplack studied Puerto Rican Spanish-English bilinguals in New York City. After analyzing lots and lots of their speech, she identified a rule that she called the "equivalence constraint," according to which code-switches are said to occur at points that won't violate a grammatical rule of either language. An example of this can be found in the following sentence. (Translations are in parentheses.)

Si tú eres puertorriqueño (if you're Puerto Rican), your father's Puerto Rican, you should at least, de vez en cuando (sometimes), you know, hablar español (speak Spanish).

In this sentence, the speaker switches languages four times, but never violates a grammatical rule of English or Spanish. This is because the speaker switches at key points in the sentence, either between clauses or to insert stand-alone modifying phrases or comments.

To illustrate with another example from English and Russian (but not from Poplack's data), most bilinguals would consider the following sentence odd or incorrect.

*Prishel a teacher.
(Arrived) a teacher.

This sentence sounds odd because it violates the rules of English. In English, the subject usually has to come before the verb. In Russian, however, it is perfectly fine and sometimes necessary for the noun to *follow* the verb. The sentence above, however, would be odd to most bilinguals because the switch comes in the middle of the phrase meaning "the teacher arrived." "*Prishel* a teacher" and "a teacher arrived" are not equivalent because the words are in different orders "*prishel* a teacher" would violate a rule against switching in the middle of a grammatical unit. And, indeed, most code-switchers probably would avoid this construction.

Being able to code-switch successfully means that the speaker has a detailed grammatical understanding of *both* languages, including what can and can't be done in *both*. This is really an impressive and significant skill. Unfortunately, code-switching tends *not* to be celebrated as the amazing linguistic resource it is. Code-switchers often think of it as something of a dirty habit, a bit like chewing gum, having a good juicy gossip session, or reading the tabloids—enjoyable and satisfying, but not something you'd do in front of your mother-in-law, for

> **• FAST FACT •**
>
> Code-mixing describes how language learners combine two languages due to *incomplete knowledge* of one or both language(s). *Code-switching*, in contrast, is common among highly proficient adults and children, and a sign of *mastery* of two languages.

instance. (And certainly not something you'd want your children to do.) This "behind-closed-doors" attitude toward code-switching exists in part because people tend to think that code-switching is the result of a lack of knowledge (when, as we've just shown, the reverse is actually true). This negative stereotyping is unfortunate as well as inaccurate.

What Does This Mean for My Children?

This take on code-switching is important when we are thinking about children learning more than one language. Most obviously, if a child grows up in a family and community where code-switching is common, she will likely code-switch herself as well, developing and fine-tuning her knowledge of how to do so through watching and engaging in conversations where switching is used with particular effects. She'll learn that code-switching is an important resource and will learn when it is appropriate (and to her advantage) to speak a certain language to particular people, and when it is appropriate to use both languages together. Code-switching in and of itself is not a problem that needs to be fixed. As a matter of fact, it is often an important and valuable resource for many children, adults, and communities as a whole.

BUT CODE-SWITCHING CAN MASK THE QUALITY AND QUANTITY OF INTERACTION IN EACH LANGUAGE

Although code-switching can be important and useful for many bilinguals, it may also present some challenges as you try to raise bilingual children who actively speak both of their languages. What's the hitch? Frequent code-switching can mask patterns of language use in homes and communities. For instance, parents who speak about equally in, say, Arabic and English, may have children who speak mostly in English with an occasional use of Arabic words. Such homes from the outside can look like ideal bilingual families with both languages in full use. Yet on closer inspection it becomes clear that English is taking over more and more of the family and community language space. If parents and children are frequently code-switching, there is no safe space for each language to develop. For example, if code-switching is the norm, children might not have the opportunity or incentive to use their weaker language, or try out new vocabulary words or grammar constructions. From the parents' perspective, frequent code-switching makes it difficult to keep tabs on the total quantity and quality of use of each language.

Parents need to keep a careful eye on language use patterns. While the combined use of two languages is alone not a problem, it can make it difficult to see what's really happening in the home in terms of the balance of quantity and quality of each language. To illustrate, let's return briefly to Oslo to consider some of the ways in which language use of two languages simultaneously can play out in family interactions. Remember that Elizabeth Lanza studied two Norwegian-American families with two-year-old children. Each family had a stated family language policy when Lanza started her research, and, indeed, this is one of the reasons she chose to work with them: In each family, the parents reported trying to stick to the one-parent–one-language rule.

However, as happens all too often with research, what Lanza thought would be perfectly comparable families actually turned out to be rather different. In actuality, each of the families had a different language style. One family adhered to the one-parent–one-language rule by strictly separating the languages so that the child always heard English from the mother and Norwegian from the father. The other family was much more lax (or relaxed, depending on your perspective), frequently alternating between English and Norwegian in the same setting. For example, during one family dinner conversation that Lanza recorded, the mother moved back and forth between Norwegian and English.

These family language style differences were most apparent when Lanza looked at how the parents reacted when the children spoke the non-target language (for instance, when the child spoke English with the Norwegian father or Norwegian with the American mother). Lanza was able to identify five main strategies that ranged from simply pretending not to understand the child to code-switching themselves.

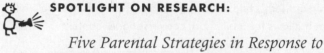

SPOTLIGHT ON RESEARCH:

Five Parental Strategies in Response to Children's Use of the Non-Target Language

- ***Pretend not to understand:*** When the child says something in the other language, the parent asks for clarification by saying, "I don't understand" or "Say that again." The child has to find a way to get her parent to understand, and this is most successfully done by using the other language. This strategy tries to enforce a *monolingual* conversational context.

- ***Make a guess in the other language:*** The parent guesses what the child might have said in the other language (for example, if the mother asks, "What does the dog want?" and the child responds, "ben," the mother might then say, "A bone?"). This strategy usually takes the form of a yes/no question (such as "Do you want ___?") and shows a preference for one of the languages over the other.

- ***Repeat in the other language:*** The parent repeats exactly what the child has said, but translates it into the language that the parent prefers the child to use. This sort of repetition strategy does not take the form of a question (unlike the guessing strategy above), and so it does not call for a response from the child. It can be said to be more bilingual in orientation in that the child is not pushed to change what he or she originally said.

- **Move-on:** The parent continues the conversation and shows an understanding of what the child has said in the other language. The parent accepts contributions from the child in either language without showing a preference or asking the child to modify the original utterance.
- **Code-switch:** The parent either incorporates the child's word(s) from the other language or changes languages on a larger scale, producing the next sentence entirely in the other language. This is the most bilingual of all the strategies because the parent follows the child's lead. In essence, this strategy says to the child, "Okay, if you'd rather use Norwegian, that's fine. I will, too." This strategy is the most *bilingual* in that it allows for a bilingual conversational context.

And the result of these different practices? One-language strategies, like pretending not to understand Norwegian and helping the child by making guesses in English, led to the child developing stronger English skills. The message for parents from this important study is twofold. First, Lanza's work suggests that children's competency levels will be higher if parents use more one-language strategies (for example, pretending not to understand or making a guess in the target language). This is especially the case when one language is in a weaker position. (In Norway, English was the weaker language; in the United States, the non-English language would likely need extra support.) Obviously, how much to push the one-language strategy (or for instance, how long or how well you can pretend not to understand the non-target language) varies greatly across families and is a personal decision, but the Oslo work suggests that to the extent feasible, these one-language strategies are worth a try. Of course, promoting the use of the target language likely has some costs (e.g., occasional frustration for both child and parent, lack of complete understanding at times), and these costs will play out differently in each family. However, such a strategy can also be quite effective in promoting the weaker language.

Second, the Oslo work reminds parents that even explicitly stated and agreed-upon family language policies (for example, one parent to one language) often play out very differently when we look at micro-level interactions in detail. These rather small differences add up over the hundreds of interactions that occur each day and have an overall impact on children's language learning opportunities. Parents may want to consider these micro moves of day-to-day conversation carefully. This requires quite an investment of time and energy (not to mention that family members might find it a bit strange!). Still, it can be a fun research project and a way of raising language awareness among family members who share the same goals.

EXERCISE:

A Mini-Research Project—
Assessing Family Language Use

With the permission of family members, record some of the following situations of language use (about ten minutes each):
- A family dinner conversation
- A play session between mother and child
- A play session between father and child
- A family project where everyone is working together (for example, cooking, cleaning up, giving the dog a bath)

Part One: How much input is the child getting in each language?

1. Fast-forward to about minute five of the recording.
2. Write down the first five to ten sentences used by the parent(s).
3. Tally the number of words in each language, and calculate the percentages (for example, No. of English words/Total No. of Words).
4. Do an analysis of the kinds of words and phrases that are being used in each language. Are some sorts of words more likely to appear

in certain languages (for example, in some houses all of the food words are in English)?

5. Prepare to be surprised: When Kendall did this in her own home, she was astonished to learn how many English words she was using when she *thought* she was speaking only Spanish!

Part Two: What might code-switching be
covering up?

1. Go to about the middle of the tape and listen for code-switches. Count how many times people change language.

2. Then note who is doing the switching. (For example, is it the parent or child?)

3. Also note, which way switching is going? (To English? To another language?)

4. What is the reaction to the switch? (For example, does conversation carry on in the switched language or the original one? Does everyone accommodate the switch?)

5. What patterns do you see? (For example, in some homes, when the child switches to English, parents follow the switch to English as well.)

(Blank worksheets are at the back of the book.)

There are some advantages in trying to promote clear boundaries between languages—at least some of the time. (Not the least of which is that it can simplify life!) And having each parent focus on providing input and interaction in a different language may be one way to keep the balance between the two languages roughly equal (assuming of course that parents are equally involved in caretaking).

At the same time, parents should be realistic. Very strict, non-stop, 24/7 separation of languages is probably not possible for most families. Indeed, in our thirty combined years of experience, we've never met any families who have kept the languages 100 percent separate 100 percent of the time. Perfect separation is certainly not the *only* or even the *best* way to raise a child bilingually. (In fact, there are lots of documented cases of children from one-parent–one-language homes who understand two languages very well, but speak only one.) In light of all this, we stress that parents should focus on the overall goal of ensuring that children receive input and interaction in sufficient quantities in each language, and not worry about maintaining overly strict separation each and every instant of their waking lives.

 QUICK TIP: Is Code-Switching Masking a Problem?

Families who frequently code-switch should make time to think about quality and quantity of exposure to and interaction in each language. They need not worry, however, that the languages must always be kept separate each and every moment.

LANGUAGE IS LEARNED IN CONTEXT (AND COMPETENCY LEVELS REFLECT THIS)

A crucial final point to consider here concerns how we tend to think about bilingualism as an abstract concept versus the flesh-and-blood reality of living, breathing, everyday bilingual children and adults. Many people (both monolinguals and bilinguals are included here) think of bilingualism in somewhat idealized terms and to define a bilingual as someone who is equally balanced in both languages. For most of us, a "true" bilingual is someone who can pass as a native monolingual speaker in a wide range of contexts in each of the two different languages. In other words, in most people's minds, a bilingual is much like two monolinguals in one body. Yet this true bilingual is largely a myth, or at least nearly nonexistent in relation to the number of people who live and function perfectly well in bilingual environments.

To illustrate this point, whenever we conduct a workshop with a group of parents, we ask them if they personally know any equally balanced bilinguals. Typically, about half of the group of thirty raises their hands. We then ask them to think more carefully about these "balanced bilinguals" and then ask themselves if the bilinguals could, for instance:

1. Read a newspaper article in both languages equally quickly and with equal understanding? *(About half of the hands fall right away.)*
2. Tell a joke or give a toast that was equally funny in both languages? *(A few more fall.)*
3. Explain to their child in both languages why it is not okay to, say, bop their friend Theodora on the head? *(One or two fall after a moment of consideration.)*

By the time we get to question 3, we typically have either no remaining hands up or one or two at the very most. When we ask that parent to explain who the bilingual person is (sometimes it is in fact themselves), they unfailingly describe someone with a very balanced and somewhat unusual background—for instance, one young mother named Anna explained that her parents were language teachers bilingual in Spanish and English, that her extended family also spoke both languages, that she had lived for extended periods of time in Puerto Rico, Argentina, and the United States, that she had attended school in each language for a significant number of years, and now had a Spanish partner.

Anna had become a balanced bilingual because her contexts for learning and using the language were balanced. The point to remember from this anecdote is that all children (and adults) learn languages *in context*. By this we mean that children learn how to do things with a language (like how to tell stories, describe pictures, ask for specific foods, understand written texts, write letters to pen pals, etc.), by *doing* those things in that language. One major reason why there are so few true or balanced bilinguals is because most people do not do everything in both languages. For instance, many bilinguals who have grown up in the United States have had the experience of speaking their heritage languages (for example, Spanish, Russian, Korean, Tagalog) at home and perhaps in their neighborhoods, but they have used English for schooling, and then later in their work lives. It's hardly surprising that these folks excel at telling stories and jokes and discussing personal issues in, for instance, Spanish, Russian, Korean, or Tagalog, but might be hardpressed to write formal memos or give technical presentations in those languages. (On the other hand, these formal tasks probably would be accomplished with relative ease in English.) This is perfectly normal and predictable. People learn to do specific things with language when they are regularly required to do those things in that language.

Learning, knowing, and using two languages inevitably impacts one's point of view, how one expresses oneself, and how one processes and uses language. Most bilinguals note that this can at times be a little frustrating. Even when bilinguals are speaking with monolinguals, for instance, they rarely manage to completely deactivate the language they are not speaking at the time, which occasionally leads to small amounts of interference in language production or perception. Still, the cognitive, social, academic, and other potential benefits of bilingualism far outweigh these drawbacks. (We've never met someone who wished they spoke *fewer* languages!) Indeed, as discussed in chapter 1, the effects of bilingualism on children's cognitive functioning often constitute a major advantage! In other words, the fact that a bilingual isn't simply two monolinguals in one body is a good thing in many respects.

Language Learning Is a Lifelong Process

Language learning of all sorts is a process. Competency levels naturally ebb and flow depending on one's environment and the interactional demands of that environment. In our research with parents, we've often found that they tend to think of bilingualism as a finite and fixed goal. Their aim is to get their kids there, to meet that major milestone, congratulate them (and themselves) and then move on to the next goal. Unfortunately, language learning is not like other developmental milestones such as crawling, being toilet trained, and graduating from kindergarten—it's a lifelong process. The downside of this, of course, is that it's never done the way that toilet training can be checked off the list once and for all once it's mastered (at least that's what we've been told about that area of development).

Parents should probably always be keeping an eye on the *quantity* and *quality* of exposure and interaction their children experience in both languages. Language learning happens best through natural or naturalistic everyday interactions, and these should be sufficient not only in terms of how often they happen, but also in terms of the range of topics and types of language required in them. Input and interaction should also be appropriately demanding for the child's academic and intellectual level. So, for instance, those baby board books in French may have been appropriate at age one or two, but they need to be updated as the child grows. Literacy skills do transfer, to a certain extent, from one language to another, but if academic language proficiency is desired, it has to be worked on continuously.

Certainly, there are some pluses here as well. For one thing, because language learning is a lifelong process, it is never too late to start, as we said in chapter 4. In addition, you can (and should) relax a bit and take a longer, more lifetime-oriented view of language learning, and not be too anxious if your children don't immediately begin using one or the other language. As a case in point, we recently attended a conference in (officially bilingual) Canada, where a famous researcher was talking about a study he had done with a group of bilingual families in Europe. The parents spoke two languages (German and French) to the children from birth, and the researcher visited each month to audiotape the children. After about six months, he was finding that the children spoke only German to their parents. (They lived in a German-speaking community.) The researcher was disillusioned and disappointed to the extent that he wanted to cut his losses, so to speak, and abandon his research project. The parents, however, were heartbroken at the prospect of being dumped and wanted to continue. To appease them (but not because he hoped to find or see anything different), he continued with his monthly tapings. Lo and behold, after nearly a year of speaking only German (and with little change in their environment), the children began to speak more and more French!

WRAP UP

The take-home message here, then, is that parents should not give up if results of their efforts are not immediately evident in their children's language. They might want to be aware of alternative approaches they can take, but they should keep in mind that children may persist in speaking their dominant language for many months at a time without it necessarily meaning that the children are not taking in the second language around them. Our own children often respond in English even when addressed in their second language. They also frequently mix their two languages. In fact, Kendall's son's favorite statement at the time of writing was, "The wheels *no van*" (The wheels don't go), In relation to his plastic Hot Wheels car toothbrush. He was absolutely right: The wheels were not real, movable wheels. He was also right, in a manner of speaking, in his use of English and Spanish together to make a meaningful, coherent, and (in many ways) grammatical sentence. And long before the time he gets some real hot wheels (heaven forbid!), he'll be able to describe what they do in full, vivid sentences in both languages.

POINTS TO REMEMBER:

- Don't worry if your child mixes languages—language mixing is a normal (and short-lived) part of bilingual development. Your child will sort this out on her own.
- Trust your child is not confused—she may not know that she's using two languages, but there's plenty of evidence to suggest that she has two linguistic systems and is very quickly learning the rules of when to use which language.
- Know the difference between language mixing and code-switching when evaluating your child's language use. If your child lives in a home where code-switching is common, she'll learn to code-switch herself.
- Keep a careful eye on quantity and quality of input and interaction in each language, especially if code-switching is the norm in your family—minority languages may need extra support.
- Set realistic expectations for your young learner—there are no perfect bilinguals in the world, and remember that language learning is a lifelong process—it's never done, and it's also never too late to start!

CHAPTER 10

What Do I Need to Know about Language Delay, So-Called Expert Advice, Special Needs, and My Child's Apparent Lack of Progress?

While bilingualism is becoming more valuable—and more valued—in countries like the United States, the fact remains that monolingualism is still very much the norm. This means that most parents trying to raise bilingual children will run into at least a few skeptics, critics, and problems along the way, and may even have to overcome some of their own internal doubts.

WHAT IF YOU CAN'T SHAKE THE NAGGING WORRY THAT YOUR CHILDREN WILL BE DELAYED BY BILINGUALISM?

Children seem to be designed to learn languages and can learn one language, two, or even three with ease in the right environment. All children take about five years to become fully fluent in their first languages, and researchers have found that children learning two languages can become fully fluent in both by about five years as well—especially if they have received approximately equal amounts of input and interaction opportunities in each. Bilingual children also hit the same linguistic milestones and at about the same time as monolingual children. Just as there is a lot of individual variation across monolingual children, there is also great variation across bilingual children. But on the average, bilinguals and monolinguals enter the one-word and two-word stages, for instance, at around the same time.

 SPOTLIGHT ON RESEARCH:

Does Learning Two Languages Result in Language Delay?

In order to answer this question, in 1992, a group of researchers compared two groups of children ranging in age from eight months to two and a half years. One group was acquiring English only, while the other was acquiring English and Spanish simultaneously. The researchers (Pearson, Fernandez, and Oller) asked parents to complete a questionnaire about the words that their children could say. This vocabulary checklist was given in English only to the English monolingual families, and in two versions (one in English and one in Spanish) to the bilingual families.

Their findings revealed that the bilingual children acquired their two languages on the same timetable as the monolingual English children. For instance, both groups showed similar rates of progress over time, and both groups went through the vocabulary spurt at around two years of age, when children begin to say many new words in a short period of time. Finally, when the bilingual children's Spanish and English words were added together into one list, the total size of their vocabularies was similar to that of the monolingual children.

Initially smaller vocabulary in each language at an early age is typically overcome by age four or five, so that by the time bilingual children go to school, their vocabularies in each language are equal to or greater than those of their monolingual peers. (And, of course, when you add both of their vocabularies together at that age, they are way beyond their monolingual peers!) So parents should rest assured that even though their children may go through a phase in which they appear to have slightly smaller vocabularies than their monolingual peers (and even though this may be a source of commentary from monolingual adults), by the time bilingual children go to school, their vocabularies have caught up with and often exceeded those of monolingual children.

It's only natural for us to want to make sure that our children are on track. In this book we've tried to give you a sense of the wide range of normal variation that exists in the process of language development, and by so doing, we hope we've also helped to reduce (or prevent!) whatever anxiety you might feel about comparing your developing bilingual child with his or her monolingual peers. Much of the anxiety we've witnessed among parents has turned out to be unnecessary, so we're especially happy if we can save you any headaches or sleepless nights!

Language learning is often used in the assessment of overall developmental progress and to screen for developmental problems such as autism or learning disorders. It is important for all parents to pay attention to apparent language delays, just as they should when their children fail to meet any (and especially, multiple) milestones. (More on language learning and special needs children later in this chapter.) Speech-language pathologists are trained to help children who are having some trouble with language and who could benefit from early intervention. Bilingual speech specialists exist in most areas of the United States, so if there is more than one language in the picture, parents should know they have the option of seeking out a bilingual therapist. It's the job of these trained professionals to look out for problems and to work to ensure that

• FAST FACT •

Bilingualism does not *cause* speech delay or any other developmental disorder.

children with language delays of various kinds—whether they are monolingual, bilingual, or multilingual—achieve academic success.

The take-home point here is that speech and developmental problems, some of which are genetically based, are *not* the result of bilingualism itself (although the results of such problems will, of course, be evident in both languages). Some children begin talking early and some begin talking much later for a wide variety of reasons, but learning two languages generally won't delay (or speed up) that pace. Parents of all children should be vigilant and feel free to consult with the professionals that many school districts provide, but they certainly shouldn't worry that an underlying problem is *caused* by exposure to two languages.

WHAT IF YOUR CHILD DOESN'T SEEM TO MAKE PROGRESS IN THE SECOND LANGUAGE?

Some children flourish immediately in a second language environment whether it be their English as second language class at public school, their Swedish mother-child play group, or their Arabic classes on the weekend. Others take more time. Each child—and each learning environment—is unique and it is impossible to predict individual learning trajectories. Some children make slow and steady progress, while others might take off like a rocket only to plateau after a few months and seem stuck at the same stage. Others might remain silent and not say a word in their second language for many months. The important thing is to stick with the second language and don't get discouraged.

This can be hard going at times. For instance, Eunjin is a doctoral student in linguistics and knew that language learning didn't happen over night. However, she still found this an anxious time with her son, Christian. She explained: "My husband, Suk-Young, and I only spoke Korean to Christian at home. We really wanted to support his Korean. However, when Christian started school when he was about twenty-two months old, I wasn't so sure about keeping his heritage language and got nervous. At that time he understood Korean very well and responded quickly to Korean speakers. But at school, he didn't know what to do. Looking at him standing at the corner of the classroom with a puzzled face just crushed my heart. He cried the first couple of days at school—as any other kids do—but I thought that he maybe cried more because he could not understand what was going on around him in this English-speaking environment. I thought he would catch up in a week, but it didn't happen that fast. Now, after going to school for six months, he can definitely communicate with his teachers. But still he is a little behind other English-speaking kids in his class."

Christian was experiencing what is sometimes known as the *silent period*. Although it may not seem like it from the outside, this is an active time for the child. During the silent period, learners are acquiring sounds and sound combinations, word boundaries, grammar and vocabulary, and observing how to use the language appropriately in different situations. The child is a *dynamic* silent observer during this time and is actively learning to create meaning in a new language. They're quiet, but they're thinking!

 SPOTLIGHT ON RESEARCH:

The Silent Period

In a 1998 study, Magdalena Krupa-Kwiatkowski describes the first three months after she and her six-year-old son arrived in the United States from Poland. Her son, Martin, outgoing and talkative in Poland, was quiet and withdrawn, initiating very few interactions with peers during play. For the first three months in the United States, Martin used mediation and attention-getting strategies to interact with other children and adults. By mediation strategies, we mean that Martin would encourage an adult to intervene between himself and other children. For example, he would ask his mother to walk him upstairs where the other children were playing, so that she could help him be included in the play. Attention-getting strategies, on the other hand, are meant to prolong peer interaction. Martin, for instance, would imitate *The Benny Hill Show* by running out of a room in different costumes to entertain the other children. Both strategies allowed increased playtime with other children without extensive use of English.

Martin's language use varied during his silent period depending on the surroundings and people. In the beginning, he was only comfortable approaching children who were also beginning-level English learners. As he grew more relaxed in his new environment, his experimentation with English gradually increased. He practiced both alone and with others, often repeating a word or phrase over and over again. He asked for confirmation from other Polish speakers about word and sentence meaning. Then, after about three months in the United States, Martin began to feel confident enough to approach children and adults in English, respond to questions and commands, and express new thoughts. After three months, his silent period ended when he "started speaking English profusely."

Parents should not worry overly much if their child experiences a silent period for the first few months, especially if they are in an immersion environment where they are expected to use a new language. It's normal. Pushing or pressuring children will not speed things up and can potentially exacerbate their anxiety. Allow your children to begin speaking in the second language when they are ready and comfortable. If your child has adequate quantity and quality of exposure, she will begin to speak soon, and once she does you may wish for a few quiet moments!

After the silent period, your child may make rapid progress, and then suddenly stall. Language learning is rarely an even and steady process. In some cases, learners will seem to get stuck, continuing, for instance, to make the same sorts of mistakes (for example, "He walk to school") for many months. This is called *plateau* or *stabilization* and may be the result of learners not being sufficiently challenged beyond their ability.

Getting past a plateau sometimes takes intervention and action. For instance, you may need to increase exposure and interaction with speakers of the target language. To get past a plateau, learners sometimes need to pay some explicit attention to language. Language games are an excellent way to encourage experimentation with the language and call attention to particular language forms. For example, if a child plateaus and applies the *-ed* past tense suffix to all verbs, saying "goed" instead of "went," you might try modeling the correct form, then make a game of reporting all the places you both visited that day ("First we went to the post office, then we ate lunch, and we went to the bank...."). This sort of past tense activity is also something you could potentially do during circle time in a language-based playgroup. Basically, you need to draw the child's attention to where they are continually making errors. Research in second language acquisition suggests that when children focus on grammar while also paying attention to meaning, long-lasting learning is more likely to occur. So, draw your child's attention to a verb form, not in a boring second language drill, but in an engaging activity like looking at photographs on a camera phone of what they did that morning. Ask your child in the second or weaker language where they went (to elicit "We went to...."), and immediately after that, return to the fun activity of scrolling through the photos.

Whether children are silent or stalled, these are normal twists, turns, and potholes on the sometimes bumpy road of language learning. Your reaction and support in these moments will help to smooth the ride. Patience, assistance, and encouragement will all contribute to the positive outcome of overcoming these obstacles.

WHAT IF YOUR DOCTOR OR TEACHER RECOMMENDS YOU DROP THE BILINGUALISM THING?

Just about every parent who is raising young children in more than one language has at least one (and lots of times, more than one) story of when their child's teacher (or day-care administrator, pediatrician, school nurse, etc.) told them that it would be better for their child if they just spoke one language. Our favorite is from our Spanish friend, Cristina. Cristina and her husband have two young boys. The boys heard mostly Spanish at home, and so, when they started attending their university-based day-care center at about two years of age, this was a big change for them. Both boys had a bit of catching up to do in terms of English-language learning relative to their English monolingual peers. For instance, one of the teachers noted one boy's English vocabulary was smaller than that of his peers.

Any guesses as to the day-care teachers' suggestions? More than once it was suggested to Cristina that she should start using more English at home with her boys. This story is instructive because this day-care center is a nationally certified and high-quality center on just about every measure. The waiting list is huge. Many of the children come from bilingual households. The staff are well trained and caring and make every effort to promote diversity. However, they missed a beat on this one. Now it just so happens that Cristina is an internationally known researcher on bilingualism and second language acquisition. So, as you can imagine, she didn't take being told to bring up her children monolingually lying down. Rather, she asked for a meeting with the day care's teachers and administrators, in which she explained the benefits of bilingualism and the importance of maintaining Spanish at home. She also held information sessions for both staff and fellow parents. The school got the picture and stopped making this recommendation to parents.

But what about other parents who face similar recommendations, but who don't happen to have the same academic credentials?

First, you need to recognize that even though Cristina has a Ph.D. in applied linguistics, this was still stressful and unpleasant for her. Even if you are perfectly confident that you are doing the right thing as a parent, *no one* likes being asked to justify their personal choices. Second, you need to try not to personalize things like this, but instead recognize that this sort of monolingual bias is unfortunately widespread in much of the United States (and elsewhere). As a country, the United States has a long history of being skeptical and suspicious about bilingualism. Even though there is no sound research supporting these concerns, in many ways, we're still living in the shadow of these outdated notions. Keep in mind that in most countries, the majority of parents would be delighted for their children to have the chance to be bilingual.

Once you're calm enough, it's worth scheduling a time to discuss your child's development in detail with the experts who are suggesting you change your ways. Ask them, for instance, to identify what exactly their concerns are in relation to your child, and how those concerns relate to language. They may not be connected at all. For instance, perhaps your child's teacher notices that your child is having difficulty making friends. She attributes this to your child's lack of perfect English skills. However, as a parent, you may know your child interacts perfectly well with English-speaking peers in other contexts (and has lots of friends in the neighborhood), but prefers to do many activities alone. In this case, the so-called problem here isn't related to language at all.

It might also be necessary to explain clearly what you are doing and why (skim chapters 1 and 6 for pointers beforehand). You might even loan them a copy of this book. Hopefully, by providing more information you can get the person expressing a concern to at least understand why you are doing what you are doing, and if possible, even to begin to share your goals.

Lastly, you need to get ready to simply ignore such well-meaning suggestions and remarks, the same way you would ignore a good bit of that well-intentioned advice about how to get your baby to sleep through the night or to stop using a pacifier: You knew the recommender meant well, but their suggestion wasn't right for your family. After reading this book, you should have a clear sense of the extensive scientific research out there on language learning among children. And of course, you are the best expert on what's best for your child. So follow your own lead!

WHAT IF YOUR CHILD HAS SPECIAL NEEDS?

If your child has special needs you probably already know that there are many books for parents on how to identify and manage different types of needs as well as how these developmental differences potentially impact first language development. We suggest you consult these extensive resources, and we have provided a list of some of the books in these areas on our Web site (www.thebilingualedge.com).

Special needs range from hearing problems to autism, attention-deficit disorder, and specific language impairment (SLI). (Each affects less than 10 percent of all families in the United States.) While these are very different disorders, what seems quite common across all of them is this: As soon as a potential problem is identified, pediatricians, teachers, and speech therapists often recommend that the whole family switch to the majority language (English in the United States) and leave bilingualism behind. Their reasoning is that children with special needs have enough challenges as it is and that they would have an easier time coping with demands in just *one* language. From their perspective, mastering two languages presents an extra and unnecessary burden for a child who is already struggling.

While this point has some intuitive appeal, we also believe it is important to keep in mind that *very* little research has been conducted on this issue. It is also important to evaluate *any* recommendation with respect to the wider context of the child's life.

Doctors and speech therapists are trained to look narrowly and in detail at the development of particular behaviors and skills. They are less trained (and have much less time) to consider the child's family and social and emotional life. And for many children, there are some significant costs associated with switching entirely to English. For instance, in some homes, switching to English means children lose the ability to communicate with their non-English-speaking grandparents, who can be a major source of love and emotional support. In other homes, switching to English has costs for the other children in the family as they lose the opportunity to become bilingual if parents enforce an English-only policy. In still others, parents themselves are not proficient in English. As linguistically rich input is important for all children, recommending that parents speak their weaker second language hardly seems ideal for a child with special needs. It's important that parents consider all of the different pieces of the puzzle together—and perhaps *with* your pediatrician or speech therapist—rather than immediately moving to English only.

Parents of children with special needs should also keep in mind that in most of the world, *all* children—those with special needs and those without—routinely become bilingual, or even trilingual. And in many homes—inside the United States and around the world—monolingualism is simply not an option. Lastly, in deciding what's best for your family, it's helpful to keep in mind that as the needs of every child are different, so too is the particular social, cultural, and linguistic environment of each child. There's no one-size-fits-all solution here, and unfortunately there is not a lot of research available on children with special needs learning more than one language from an early age. That being said, though, most researchers and practitioners would agree that intensive and interactive exposure to appropriately complex language is important and beneficial for *all* children and that bilingualism does not need to fall by the wayside. It will probably also be important to seek out professional assistance from speech-language pathologists, psychologists, or pediatricians who are familiar with the specific learning processes of bilingual children. At the same time, remember *you* are also an expert on your child's needs!

EXERCISE:

Considering the Big Picture for Your Special Needs Child

1. What are the communication needs of your child? What purposes or benefits come from introducing a second language or maintaining both of her languages?
2. How will you, your child, your family, community, and school operate in one or both languages?
3. What are your child's strengths?
4. What are your child's weaknesses?
5. What sorts of support are there for both languages?
6. What will be lost if you switch to English (for example, relationships with grandparents)?
7. What will be gained by promoting bilingualism?
8. How much time, energy, and additional resources will be needed to provide a language-rich environment in both languages?
9. How does your child seem to respond to/handle two languages?

Adapted from Toppelberg et al., 1999

WRAP UP

What happens in the real world of parenting is rarely predictable and often demanding. Challenges to your language plans are going to present themselves sooner or later *in every family.* Keeping the benefits of bilingualism in mind along with a long-term perspective, paying attention to your children's needs, and being flexible in your approach are important strategies as you forge the road ahead.

POINTS TO REMEMBER:

- The various life challenges we face need not be insurmountable roadblocks to bilingualism.
- Remember, it's normal for children to go through a silent period when first immersed in a second language.
- Keep in mind the benefits of bilingualism when making decisions about whether to continue on the road to bilingualism.

CHAPTER 11

What Do I Need to Know About Trilingualism and Dialects?

In countries like the United States and Britain, speaking only one language, or being *monolingual*, has long been seen as the regular, healthy, and so-called "normal way of being." These perceptions are gradually changing as families become more culturally and linguistically diverse—and aware. Nevertheless, for many, command of two or three or more languages—known as *bilingualism, trilingualism, or multilingualism*—is still seen as elite, exotic, unusual, or even abnormal. While this perception is widespread, it's important to keep in mind that children from all over the world routinely become competent speakers of several languages simply by being regularly exposed to—and interacting in—multiple languages. For most of the world, bilingualism or trilingualism is typical, uneventful, and indeed, very much the norm!

LEARNING MORE THAN TWO LANGUAGES
IS PERFECTLY POSSIBLE

With more than six thousand languages coexisting in fewer than two hundred countries, it makes sense that multilingualism is the world norm. For example, 90 percent of the population in African countries such as Nigeria use more than one language regularly. Switzerland has three official languages (French, German, and Italian) and a fourth national language (Romansch). Many countries in Asia are also multilingual. Children from Africa, Europe, and Asia are not born better equipped than children in the United States to learn more than one language—their brains or mouths are not innately more capable. Instead these children simply become multilingual because they have a reason to. By "reason" we mean both the *opportunity* and *incentive* to interact in more than one language. For instance, children in Luxembourg grow up speaking Luxembourgish at home with their parents and family. When they enter school, they start learning German, French, and English. Likewise, a child in Singapore might speak only Tamil until entering school, where English is the primary language of instruction, and then later on, study Chinese as an additional language.

Children from these countries are not freak whiz kids. They are normal, everyday children. Likewise, your bilingual or trilingual children will be in good company. Most children of the world grow up surrounded by more than one language! The fact that so many of the world's children learn multiple languages with ease clearly demonstrates that three or more languages pose no particular problem, risk, or threat for the vast majority of children. Indeed, from a scientific perspective, there are millions of data points (in the form of healthy, happy multilinguals) to support this claim! So wherever they live, parents aiming to raise multilingual children should take heart that knowing several languages, in many respects, is nothing special, certainly not risky, and well within the realm of possibility.

Acquiring three or more languages is similar in many ways to acquiring two. For instance, children learning three languages from birth begin to say their first words around the same time as monolingual children. They also begin to combine those first words into two-word sentences like "no sleeping" or "wanna cookie" at around the same time.

SPOTLIGHT ON RESEARCH:

Multilingual Children— Some Smashing Success Stories!

Many language researchers have studied their own children in close detail. By collecting lots of detailed data (that only a parent-researcher could get!) such as systematic weekly audio and video recordings of everyday home moments, these researchers have added greatly to our understanding of multilingualism. A few examples:

• In 2000, Jean-Marc DeWaele studied his own daughter, Livia, and her acquisition of four languages—Dutch, French, English, and Urdu. She was exposed to Dutch by her mother, French by her father, English in the greater London community, and Urdu by her nanny from five months to two and half years of age. When Livia was a toddler, DeWaele reported hearing his daughter playing with her dolls and repeating rhymes and phrases in all four languages in her nursery. More recently, DeWaele has reported that Livia, now an eight-year-old, uses only three of the original four languages because she no longer interacts with her Urdu-speaking babysitter.

• In 2006, Madalena Cruz-Ferreira published a book about raising her children in Swedish, Portuguese, and English. She spoke only Portuguese with the children, her husband spoke Swedish, and the family moved to Singapore, where English was commonly used outside of the home and was the language of schooling. When the youngest son entered school in Singapore, the three children began to speak English among themselves, but they still used Portuguese and Swedish with their parents.

In some of the most successful cases, the family language goal of multilingualism is supported by very particular circumstances. For instance, one European family we know successfully introduced four languages to their daughter, Nora. The family lived in Stockholm but the mother (Maria) was from Spain and the father (Fabrizio) from Italy. Because Maria and Fabrizio had met and married in London while completing their university studies, the language of their relationship had always been English. Upon graduation they moved to Stockholm for work. When their daughter, Nora, was born a few years later, Maria always spoke with her in Spanish, while Fabrizio spoke to Nora in Italian. The parents continued to use English with each other in the home. Nora began to learn Swedish when she enrolled in the state-run day-care system at age one. Nora is now seven and fully proficient in all four languages.

To be fair to the rest of us, we should note that Maria and Fabrizio had some big advantages and unusually good circumstances in their favor. For instance, both parents are trilingual themselves (in Spanish, Italian, and English), so no one ever felt left out; and both of their extended families lived relatively close by (a short train or plane hop away), making frequent visits possible. Nora's language learning work was probably made a bit easier by the fact that the four languages were from just two language families (Spanish and Italian are Romance languages; and English and Swedish are Germanic). They also had a chatty little girl who loved to talk, so stimulating lots of interaction in and outside the home.

Still, as proud as her parents are of her, little Nora is not an exceptional genius. Rather, she happened to be born into the ideal circumstances to promote *quadra*lingualism. Most children in similar

> **• FAST FACT •**
>
> Multilingualism poses no special risks for children—acquiring three or more languages is not significantly different from acquiring either one language or two languages.

circumstances, that is, with roughly equal opportunities and incentives, would also acquire all four languages. But parents who do not find themselves in similar circumstances (for example, parents living in monolingual communities) should not conclude that multilingualism is beyond the reach of their children. They can well achieve similar goals, and indeed, it is not that difficult if some guidelines are kept in mind. While most of what we've written in this book on bilingualism applies equally well to parents (and children) who are working with more than two languages, now we will focus exclusively on the challenges and opportunities that come with multilingualism, as well as with two different dialects of the same language.

SUCCESSFULLY PROMOTING MULTILINGUALISM OFTEN REQUIRES FAMILY LANGUAGE POLICIES

Attempting to maintain just two languages can be a bit of a balancing act. This is even truer for parents who have three or more to manage. The biggest difference between promoting bilingualism and promoting trilingualism is that balancing three languages (or more) in most homes requires more management and attention from the parents. Without the right balance of language exposure, children who are learning three or more languages are likely to become passive users of one of those languages. For instance, if a trilingual English-Arabic-Persian child engages in conversations every day in English and in Persian, but only *hears* Arabic on the weekends, she'll be slow to develop an active competence in Arabic. In other words, she'll grow to *understand* Arabic but likely will not use it to communicate herself.

The challenge for parents of budding multilinguals is not so much in getting their children to start talking, but rather in ensuring that their little ones have access to roughly equal *quantity* and *quality* of input in each of the languages. By quantity we mean the amount of time in which they are exposed to each language; for instance, for a child to be a roughly balanced trilingual she would ideally spend approximately one-third of her waking hours interacting in each language. By quality we mean what types of language is she hearing, and in particular, what types of interactions is she engaging in, in each of those languages.

In multilingual countries like Nigeria and Luxembourg, this is not particularly difficult. Children consistently hear two or more languages, and regularly have opportunities to use those languages with neighbors, playmates, and parents. Multilingualism is a societal norm in these places. However, what happens when you live in a predominantly monolingual country or neighborhood? How can you ensure that your child gets the exposure and conversation opportunities she needs to become multilingual?

One possibility is implementing some fairly firm language policies in the home. For instance, the mother might speak only German with the child and the father speak only French. The third language, English, is then acquired through friends, playgroups, preschool, and in the neighborhood. These language policies, and in particular the one-parent–one-language strategy (discussed in chapter 6), are sometimes important for maintaining structure and balance in successful multilingual homes.

Another possibility is seeking out opportunities outside of the immediate family for the child to use the languages—for example, language for children classes, playgroups, and multilingual babysitters. These are all great ways to give your child a chance to use and hear the languages, but again, some points should be kept in mind.

CONSIDER YOUR GOALS RELATED
TO MULTILINGUALISM

Some questions parents wanting to raise multilingual children should ask are: What do we want, language-wise, for our children? Do we want them to be equally proficient in all three languages? Do we want them simply to have an appreciation for different languages? There are no right or wrong answers to these questions, of course, but your answers may influence how you go about finding ways for your child to hear and use the languages. For instance, Kristen and Miguel have a three-year-old son, Lucas. Both Kristen and Miguel are native speakers of English, and although they know some Spanish, they are still at the beginner level. Thinking about their own difficulties learning a second language later in life, and hoping to give their son both an appreciation for language and a head start in his education, they would like to raise their son to be as proficient in as many languages as possible. For this reason, they take Lucas, now four, to a Japanese class every Saturday and to a Spanish-only playgroup every day after school for two hours. In addition, they have a Tagalog-speaking babysitter each Tuesday. They are wondering: *Is this enough to help Lucas master all four languages?*

Kristen and Miguel are asking an important question. Our answer: If Miguel and Kristen's goal is simply to give their son an appreciation of foreign languages, an opportunity to meet other children, and a chance to learn more about topics he may, at a later age, pursue in more depth, then Kristen and Miguel's approach is just fine. *However,* if Miguel and Kirsten want Lucas to be able to speak all four languages (English, Japanese, Spanish, and Tagalog) *equally* well, then they should reconsider their approach. It is important for children to use and hear their languages in approximately equal amounts if they are to become active users of all the languages. In Lucas's case, given the amount of exposure to each, most likely he will be English dominant with some proficiency in Spanish, since those are the only two he uses on a regular basis. He will no doubt learn to understand some Japanese and Tagalog, and maybe even say a few key words, but the chances are slim that he will be able to speak them as well as he does English or Spanish.

Reflecting on the issue, Kristen and Miguel decided that it is important to them that Lucas become as close as possible to equally proficient in English and Spanish. Given their goals, and since neither parent is a proficient speaker of Spanish, an argument could be made for dropping the Japanese language class and the Tagalog-speaking babysitter in favor of a Spanish language class and a Spanish-speaking babysitter—in other words, to optimize exposure to the second language (Spanish) and to develop as much proficiency as possible in that second language before introducing a third. This may give their son the best odds at high-competency levels in both Spanish and English, since the family is putting their full weight, energy, and effort behind it—rather than spreading their energy and efforts thin over a wider array of languages.

But of course, this approach would not be appropriate for all families. If the parents live in an environment that supports multilingualism (for example, a multilingual country, neighborhood, or family), then it is not necessary to concentrate on only two. As we stressed at the outset of this chapter, multilingualism is common and does not overly tax the child's brain. Similarly, if parents are not set on having their children be equally proficient in more than two languages, it is not necessary to replace the Tagalog-speaking babysitter or drop the Japanese language class. However, they will want to consider extra input to "up" the level of exposure of these two weaker languages.

So learning more than two languages presents some special challenges, and parents should think carefully about what is realistic for themselves and their children and plan accordingly. For some families (for instance, for Maria, Fabrizio, and Nora), learning three or more languages right from the start makes sense as each language has a prominent role and special place. For other families (such as Kristen, Miguel, and Lucas), sticking with two only for a while is probably a better plan. But parents of such budding bilinguals should take heart, as knowing two seems to help give children an edge in acquiring a third.

KNOWING TWO LANGUAGES MAKES IT EASIER TO LEARN A THIRD

The growing consensus among researchers is that bilingualism does not hinder the learning of an additional language in any way, but rather *supports* the acquisition of a third (or even fourth) language. Why might bilinguals have an edge over monolinguals in learning an additional language? No one is exactly sure, but it seems likely that bilinguals acquire this edge through the experience they gain in hearing, speaking, and learning more than one language. When a child learns a second language, she learns not only how to speak and use the language, but also some more global strategies about *how to learn* a language. Many of the processes and strategies that a child acquires through learning his second language can then be put to use when he is picking up his third.

 SPOTLIGHT ON RESEARCH:

Bilinguals Have an Edge in Acquiring a Third Language

In 1994, Jasone Cenoz and Jose Valencia investigated the effects of bilingual education (in Spanish and Basque) on students' learning of a third language (English). They looked at the English language achievement of 320 teenagers (aged seventeen to nineteen) and found that bilingualism (in addition to other factors such as motivation and amount of exposure) was positively associated with success in learning English. These students' bilingualism had helped them learn a third language once they got to school.

As we said in chapter 1, many researchers have suggested that bilinguals have greater cognitive flexibility and metalinguistic awareness, and that these skills give them an edge. Bilinguals might also have an edge in language learning because they already know that two words can refer to the same object (i.e., that the furry animal called a "cat" in English can also be called *chat* in French). So they are better able to manipulate words and meanings in additional languages they encounter as they grow up.

For any number of reasons, learning, knowing, and using two languages seems to provide an edge in acquiring a third. In short, some of the work that children do in learning their second language will support and strengthen their efforts to learn the third. But what if two of the three languages at play are quite similar (for example, like the different types of French spoken in France, Canada, and Haiti)? Do all of the above findings still apply? To begin to answer this question we have to consider the difference between a dialect and a language, and how this affects families.

CHOOSING WHICH DIALECT(S) TO USE IN YOUR FAMILY IS AN IMPORTANT AND PERSONAL DECISION

All languages and all dialects of the world are equally complex and equally systematic. Linguists *disagree* on a lot of things, but we all *agree* that there is no such thing as an inferior language or dialect. However, saying that two languages are *linguistically* equal is not to say that they are *socially* equal. Languages—and in particular dialects—come with different levels of social prestige. This may be a source of concern for parents, and as we

> • **FAST FACT** •
>
> Many of the skills that children develop to learn a second language can be used to learn a third.

describe below, has some important implications for bilingual parenting.

A question that parents often raise is, "How different do the two dialects have to be to count as different languages?" For instance, one family we know was trying to raise their three children (three-year-old Nicolas, and baby twin boys Sebastian and Matteo) bilingually in Italian and English. The mother, Hilary, is from Chicago and the father, Vincenzo, hails from central Italy. The entire clan now lives in Los Angeles but visits Vincenzo's Italian family each summer for several months. Vincenzo and his extended family (and indeed their whole village) on a day-to-day basis speak a dialect of Italian known as Marchegiano, which is quite distinct from standard Italian. Hilary and Vincenzo sometimes worry about which version of Italian to try to use with their sons, and also, given the quite substantial differences between the two versions of Italian, whether they actually count as different languages. In other words, if all three sons master all three varieties (standard Italian, Marchegiano, and English), are they in fact trilingual? To begin to address these issues, we have to consider the question: What do we mean by "language" anyway?

What's a Language?

How can we determine whether two varieties of speech are different languages or different dialects of the same language? One approach might be to ask a native speaker of standard Italian (say, a businesswoman from Milan) to listen to someone like Vincenzo speak in Marchegiano and then ask her to what degree she understands Vincenzo. Through this type of test, we can get a sense of whether speakers themselves consider the two languages *mutually intelligible.* Potentially then, we could say that if the two are indeed mutually intelligible, they are the same language; if they are not mutually intelligible, then they are best considered different languages.

Unfortunately, this gets messy quickly. For instance, two dialects of Chinese (Mandarin and Cantonese) are mutually *unintelligible*, but they are still considered, for political and cultural reasons, to be dialects of the same

> **• FAST FACT •**
>
> There are no agreed upon scientific criteria for distinguishing a language and a dialect.

language, rather than different languages. In turn, Serbian and Croatian are very similar, with shared vocabulary, grammar, and idiomatic expressions, but they are considered (again, for political reasons) to be separate languages rather than dialects of the same language. So what's considered a "language" and what's considered a "dialect of such and such language" is nearly always a cultural, political, or social call, and *not* a linguistic one. Mutual intelligibility tests tell us much more about what speakers *think* about the language or the speakers of the language than how linguistically similar the languages actually are.

Dialects and My Choices for My Child

So what does the above discussion have to do with parents' choices for their children? Quite a bit it turns out. First, parents should keep in mind that dialects, although they are often held in lower esteem than the standard in some ways, are not broken or error-ridden versions of the standard. All dialects, like all languages, are rule-governed and systematic. In everyday terms, this means that all dialects follow rules and patterns for signaling meaning that are logical and predictable. For example, in West Ulster English, which is spoken in Ireland, linguist Jim McCloskey says that questions like "What did you get all for Christmas?" are perfectly correct and correspond to other English versions such as "What all did you get for Christmas?" West Ulster English allows the word "all" to appear in different places in the sentence in certain situations. This is a systematic rule in West Ulster English, and not an error or mistake by a speaker of standard English.

A similar example can be found in African American Vernacular English (AAVE) and the use of "be." The PBS series *Do You Speak American?* gives the example in AAVE of "the coffee be cold." Most standard English speakers think this means "the coffee is cold (right now)." But to AAVE speakers, the sentence means "the coffee is *always* cold." "Be" here is used to mean something that happens habitually or regularly. Again, this reflects one of the grammatical rules of AAVE and is not a mistake.

What this all means is that parents should not think that one variety of a language is corrupt or wrong. All languages and all dialects are equally complex and have rules. So, if you are a speaker of West Ulster English, African American Vernacular English, an Italian dialect such as Vincenzo's, or any other nonstandard variety of a language, this way of speaking is *not* wrong or inferior and it's perfectly valid and appropriate to use this language with your child.

Another point to keep in mind is that children don't have much trouble learning new dialects or switching between dialects. For instance, there is a lot of evidence that children who grow up speaking a dialect of Italian (like the one used by Vincenzo's family) have little difficulty with standard Italian when they begin formal schooling at age five. Similarly, children who speak Swiss German at home have little difficulty using standard German (which is actually quite different from Swiss German) when they start school. These children don't experience more difficulties at school or problems with learning because of the switch in languages or dialects. So, if you are a native speaker of a so-called nonstandard variety of a language, you should not worry that your child will have difficulties when she enters school and begins using the standard variety. Research suggests this is simply not the case.

Of course, we cannot dismiss the fact that although all languages and all dialects are equal in linguistic terms, in *social* terms, they are rarely equal. Some are considered prestigious and others are not. In practical terms, what this means is that some languages and dialects will have more materials (for example, books, songs on CDs, movies, games) than others. For example, Vincenzo and Hilary, when loading up on Italian toddler books for their babies and videos for their older child, don't have many choices. Standard Italian is about all there is. And their boys will learn standard Italian in any foreign language learning course they participate in when they are older. What's more, in Los Angeles, there is far more social support and greater opportunities to interact in standard Italian.

However, this does *not* mean that parents have to discard their own language or dialect in favor of the standard. The language of your ancestors, of your family, and of the region of the world where you're from is something to be cherished and passed on to future generations. Research suggests that your child is capable of learning more than one variety of the language. Encouraging and providing children with the opportunities to learn more than one language will enrich their lives and their language learning experiences.

WRAP UP

We hope that this chapter has given you the confidence—as well as some of the tips and techniques—to successfully introduce your child to more than two languages. With the right balance of opportunities and incentives, trilingualism is well within reach. So if the stage is set in your family, our advice is to *go for it*!

 POINTS TO REMEMBER:

- Be aware of the trilingual environment: How much is each language spoken? What quality of interaction is there in each language? You may want to revisit the Family Language Audit in chapter 6 to help you get a handle on this.
- Consider making some family language rules and try to stick with them! Successful learning of three languages or more typically requires a bit more planning.

- Balanced *bilingualism* may contribute to successful *trilingualism*. In other words, knowing two languages can make learning a third (or more) easier down the road.
- The line between a dialect and a language is hard to pin down. All language varieties are equally complex and systematic. Choose to use the variety that is important to you and your family.

CHAPTER 12

What Can I Do If My Family Disagrees, My Child Resists the Second Language, My Family Circumstances Change, or If There Are Other Problems Along the Way?

The other day as we were walking into a sandwich shop near campus, we passed a mother and her two girls, about four and six. We'd seen the family at the park a few times over the last year. They stood out because all three had long flowing reddish-blond hair, and all three spoke fluent French together. This day at the sandwich shop was different. Both girls had their hair pulled back tight into pigtails and were complaining loudly to their mother (in English), "But we hate French! Why do we have to even speak that stupid language!?"

This chapter is about parenting in the real world. By the real world, we mean the one where children don't behave exactly as we hoped or planned, where things get spilled, broken, and lost, where we lose our patience and run out of energy at times even before breakfast is finished, where our extended families are not always enthusiastic and supportive about our approach to parenting, where husbands, wives, and domestic partners of any sort, fight, disagree, or even ignore each other altogether at times, where there are long to-do lists each day, and sometimes there is not enough energy, time, or money to meet everyone's wants or needs. Being a parent is hard work. Everyone at some point wishes they had done things a bit differently at the end of the day. We all deal with tantrums, children who outright reject a second language, skeptical friends and colleagues, and spouses or extended family members who resist or undermine our efforts. We hope to give you a few real-life tips that will help you get through those little (or not-so-little) bumps; some reassurances that you can turn to when your goals seem to be slipping away; and a bit of factual ammunition you can use to quiet those skeptics you are sure to encounter in life.

WHAT IF YOUR FAMILY CAN'T AGREE?

Many families start off all on the same page in terms of language goals in the early months. When the baby hasn't said so much as "ga-ga," it's relatively easy—especially for first-time parents—to agree that she'll be the perfect bilingual child (as well as perfectly well behaved, a perfect violin player, and perfect whiz at math). At this early stage, bilingualism is a shared goal that you plan to stick with through thick and thin.

And then real life—and real parenting—kicks in, and things don't go exactly as planned. Perhaps your child starts to reject the language; maybe he is diagnosed with a learning disability or as hyperactive; perhaps she took to the second language very quickly at first, but now doesn't seem to be making much progress (and is much more interested in gymnastics now anyway). Maybe in the face of such real-life twists and turns, you and your partner disagree about the best course to take.

Or perhaps your extended family objects a few years down the line. For instance, we have a colleague, DeeDee, who is raising her son trilingually in Taiwanese (DeeDee was born in the United

States to Taiwanese parents), Croatian (her husband's native language), and English. Things were going pretty well until their family went to Croatia for the summer holidays when her son was three. Her in-laws were less than pleased when their grandson used mostly Taiwanese, his dominant language, during the visit. They had expected him to use only Croatian in Croatia. Despite everyone's best attempts—or given that he is three, perhaps *because* of those attempts—he insisted on using Taiwanese with his mother, even when he knew that no one else around him could speak that language. This put DeeDee in the uncomfortable position of either answering her son in Croatian, which tended to upset him, or answering him in Taiwanese, which was rude to her in-laws and gave the impression that she was not supportive of teaching him Croatian. The whole situation was exhausting for DeeDee and caused a fair amount of wear and tear on the relationships between DeeDee and her in-laws, DeeDee and her husband, and even her husband and his parents, who ultimately blamed him for their grandson's "deficiencies" in Croatian.

These sorts of tensions are bound to exist because language choice isn't just a practical decision—it is also an emotional one. Language has a way of intensifying whatever family issues or tensions already exist. For instance, in DeeDee's case, her in-laws were already resentful of the fact that their grandson lived far away and only visited periodically. Language—and Taiwanese in particular—became symbolic of that distance.

Such family tensions are very common as your idealized bilingual baby grows into a real child, with needs, wants, and the occasional tantrum. In some families, these tensions can even threaten to undermine bilingualism as a family goal.

Even in families where the children *are* speaking the grandparents' languages, tensions can arise across generations. For instance, Susana and her husband are from Barcelona and have lived just outside of Washington, D.C., for more than ten years. They have two boys (eight and eleven) and are doing their best to raise them bilingually. The boys attend a two-way Spanish-English immersion school and are fluent in both languages. They have lots of Spanish-speaking friends from Latin America and a regular babysitter who is from Ecuador. So far so good ... until they return to Barcelona to visit the grandparents, who can't help but wrinkle their nose at their grandchildren's Spanish. "Why can't they talk properly?" they ask. Their accents sound funny and foreign to the Barcelona family, and the children get no small amount of grief for sounding like Latin Americans instead of Europeans.

Such tensions are unfortunate, unnecessary, and ultimately undermine everyone's efforts to use the language. What's the best way for you to diffuse them? We've seen several tactics (used together or alone) that have had some success. First, remind yourself and your feuding family members that language learning is a long process and that this is just one bump in the road. It may help to explain that it's only natural for a child to acquire the accent of the region where she grows up and that the fact that she is able to communicate in the language at all is a success.

Second, remind yourself (and your fellow caregivers) of your original language learning goals and the reasons for opting for second language learning. It might also be helpful to discuss these goals with extended family members so that everyone is on the same page and working toward the same end. If grandparents insist on a certain variety of the language, perhaps they could become more active in the language learning process with weekly phone calls, letters, e-mails, or extended visits.

Third, try not to make language the battleground for playing out other issues or problems between you and your partner. Rather than play a tug-of-war over language (no one will win!), try to get to the root of the problem. Sometimes, other tensions (for example, disagreements over bedtimes or financial stresses) play out across day-to-day mundane interactions, such as what language to have a bedtime story read in. Focus on the other problems and don't let them get in the way of promoting bilingualism.

WHAT IF YOUR CHILD REJECTS OR
RESISTS THE SECOND LANGUAGE?

In the real world, despite our best efforts to make language learning fun and engaging, children *do* say things like, "I hate that language" or "No. No! NO! You can't make me speak it!" As children grow up, they rapidly become more aware of—and more susceptible to—their surroundings. By the age of two, most toddlers have a sense of what sorts of people speak what sorts of languages. And by the time they reach age five, forces outside of direct parental control—for instance, what happens at school, what their friends say and do, the messages of pop culture—start to have a major impact on your child's identity and how that identity is expressed.

Clothing is a great example. For the first few years in life, babies and young toddlers are pretty much happy to go along with their parents' choices in clothing. (If you have a child this age, enjoy all those cute outfits while you can!) As they grow (for some children this phase hits around the terrible twos, for others, a bit later), they have increasingly strong feelings about what they want to wear. In Kendall's house, a much-too-small Thomas the Tank Engine T-shirt was *de rigueur* for an entire summer. In Alison's house, Miranda insisted on wearing a stained and faded bunny rabbit motif top for the same period.

This battle over clothing in many ways parallels the struggle over language that can develop within families. Language, like clothing, has an important *functional* component (for example, communicating needs, thoughts, and desires in the case of language, or keeping us warm in the case of clothing). Language and clothing also have strong *identity* components as well. Both serve to reflect something about who we are, and at the same time, project and help create who we want to be. When we pick out a pair of high-heeled boots, for instance, we are both *reflecting* and *projecting* a certain femininity and glamour. We make similar choices with language each and every day, highlighting aspects of our identity through use of language (or languages) in particular ways (with word choice—for example, using *powder room, toilet, bathroom, loo, little girls' room*—or with language choice—for example, trying to use a tiny bit of Japanese at a local sushi place to seem more with it).

Like it or not, children become more and more independent each day, and simultaneously begin to put greater value on the opinions of people *outside* the family unit, in particular friends and peers. Their identities—and their ideas about how these identities are constructed through, for instance, particular language or clothing choices—form young and fast. As we've said before, children of all ages are *very* sensitive to what languages are "cool" or even what "cool" things are associated with what languages. At the same time, most kids don't want to stand out as different. And if learning or using a second language makes your child feel odd in some way, that's also cause for rejection.

A bittersweet example comes from our colleagues Judith and Evelyne, who are bringing their two girls up to speak both French and English. Both parents are French-English bilinguals. Their girls are lucky to both be enrolled in an excellent (lottery-based admissions) bilingual French-English middle school program in a local public school. This program is very highly rated by parents and teachers, and both girls are doing very well there. The family speaks a mixture of French and English at home, and on their regular family vacations in France and Israel, they are surrounded by French monolinguals in Evelyne's family. Imagine Evelyne's surprise when, on coming home from school one day, her younger daughter, Ava, told her she had the absolute best day in her whole life. When Evelyne asked her why, Ava explained that they had a substitute teacher in her class that day, and the teacher couldn't speak French, so the children got to speak English all day long!

Stemming a Bilingual Rebellion

There are several steps you can take to ensure that such rebellion is short-term and does not result in total abandonment of bilingualism as a goal. The first is to remind yourself that language is a personal choice and deeply connected to your child's identity. Just as it is pretty much impossible to make a four-year-old do and wear what you want her to each day, it's also extremely unlikely that you can control each word that comes out of her mouth. And just like the clothing battles, some negotiation and strategizing are in order. Perhaps the most important thing is not to allow language to become a battlefield in the first place. All children go through a period of asserting independence and defying their parents, but if at all possible, redirect those mini-battles of independence into other domains (such as clothing, toy choices, etc.). This may mean even pretending that you don't care overly much *what* your child does with language for a bit.

QUICK TIP: Avoiding Arguments About Language

Don't let language become the stage for power struggles between you and your child.

The important point here is that you can't control all of the choices your children make. You can, however, do many things to set the stage for your children to make optimal choices. So how best to set the stage? First, make a habit of talking to your child about all of the cognitive, social, cultural, and, if applicable, family benefits of bilingualism in terms they can understand. Explicitly stating different rationales from time to time should be a regular practice, as well as pointing them out when opportunities present themselves in daily life (for example, when you come across a sign in Spanish, you can point out that most English speakers—including adults—have *no clue* what that means, but that your child does!).

It's also a good idea to take a moment to ask your child (depending on her age) to explain in her own words why being bilingual is important. Such an approach allows your child to take an active role in family language planning. Let your child hypothesize a future with or without bilingualism.

EXERCISE TO DO TOGETHER WITH YOUR CHILD:

"When I Grow Up . . ."

Depending on the age and personality of your child, you may want to let your child finish one or more of the following sentences. Discuss their responses together in a way that the emphasis is on what *they* want and the advantages for them.

• Speaking a second language is important to me because . . .
• Speaking two languages helps me . . .
• I think speaking two languages is cool because . . .
• If when I grow up I speak (only) one language, I would feel (would be able to) . . .
• If when I grow up I speak two languages, I would feel (would be able to) . . .

Post reminders (for you and your child) around the house. For instance, use language and concepts that address the specific interests of your child by creating a top ten list with your child.

EXERCISE:

Top Ten Reasons Why
Speaking Two Languages Is Cool

Complete with your child, coming up with your reasons. Here is an example of a top 10 list:

10. I can have private conversations in my second language.
 9. When I told (friend, teacher, neighbor) that I speak two languages, s/he was thrilled/very impressed.
 8. I can meet more people from around the world (list of potential places where your child can meet people who speak the second language).

7. Many athletes are bilingual, like Yao Ming, Roger Federer, and Maria Sharapova.

6. I can wear and read clothes that have the print of the second language.

5. Many of my favorite celebrities like Antonio Banderas, Jennifer Lopez, or Yo-Yo Ma speak two languages, or more!

4. I can make more friends (list the names of your child's bilingual friends here).

3. After finishing school, I'll be able to get an awesome job that pays more, like _____.

2. I can help other people communicate who don't speak two languages.

1. I feel proud and good about myself.

Along with reasoning with your child and including her in the decision-making process, you might also try to make the language interesting by using things she is interested in, like media and technology, to promote the second language. Cool music, video games, and DVDs can go a long way toward enhancing the status of the second language. (See chapter 7 for more details.) Also consider whether your child has enough language role models. For instance, if he only sees adults regularly using the language, he'll be more likely to see that language as something not for him. If this is the case, try to set up playdates with speakers of that language or go to environments where you might find kids speaking that language. For instance, on many Sunday afternoons, Kendall takes Graham to a park in a Latino neighborhood very near her house. It's a slight change of scene from the park across the street where they usually go, and it also gives Graham the chance to run around and play with the slightly older Spanish-speaking boys who, like all toddlers, he looks up to. Other parents we know have made an annual trip (for example, to a family home town, to a French-speaking part of Canada, or to family-friendly destinations in

Mexico) a big incentive for language learning in their families. Planning and talking about the trip—and how children will need to speak their second language—can be a big motivator.

These suggestions may do wonders or may do next to nothing—depending on your child's age and personality. Some children are strong-willed and very independent from day one. And finding what tips work (or what combinations of different tips work at different ages) may require some experimenting. Here are some of the top tips we've heard from real parents that may work for your family.

QUICK TIPS: Parents Share What's Worked for Them

- Anna (mother of Javier, age five): "We've had the most success using reverse psychology. Javier is super strong-willed and it's actually helped at times to say to him something like 'Spanish is only for grown-ups.' Nothing makes him want to use it more than thinking he can't."

- Lucinda (stepmother of Matilda, age four): "Matilda went through phases of responding to us in English when we would use Portuguese with her. We just made a point of continuing in Portuguese. We didn't switch to English, but we didn't make a huge deal about it either and eventually the phases passed and she started using more Portuguese again."

- Fabio (father of Lorenzo, age six, and Alessandro, age four): "I decided to turn speaking Italian into a game. The three of us took turns being the 'English policeman' each day. The policeman's job was to fine whoever was caught speaking English. The penalty was ten cents into a jar. It made sticking with Italian more fun for everyone and really helped—at the end of the month we'd go out for pizza with our earnings."

- Maria (mother of Samantha, age three): "When Sammy was little I would just pretend I didn't understand English. She had to use French in order to get what she wanted. As she got older, she realized pretty fast that I understood English perfectly well. I don't ignore her requests in English now, but ask her to say it in French in the same way most parents ask their kids to say *please* or *thank you.* I've explained to her that this is our special secret language that only the two of us know, so we want to make sure we practice it."

- Cristina (mother of Jason, age nine, and Sandra, age six): "Once the kids got to be about four, we instituted a 'beeping policy.' Even though Spanish is my second language, my husband (who is from Venezuela) and I decided to make Spanish the only language of our house. Instead of nagging them, or saying, 'Speak Spanish' a hundred times a day, we 'beep' each other if we are caught speaking in English. The kids *love* to 'beep' us and shout 'BEEP!' if they catch me using English with my husband. It's made a huge difference in keeping Spanish the only language of our house."

Each of these strategies seems to have worked because—in one way or another—they make speaking the language desirable *for the child.* Each of these parents has succeeded not in forcing the language on their child, but in presenting it as an option, not a mandate. The most powerful and sustainable learning comes when *children themselves* make this choice.

WHAT ABOUT WHEN BILINGUALISM MAKES YOU OR YOUR CHILD STAND OUT?

Like it or not, monolingualism is still the norm in most of the United States. According to the U.S. Census Bureau, only 9.3 percent of Americans speak both their native language and another language fluently (compare this with 52.7 percent of Europeans!). In real life, these statistics mean that there will probably be times when you or your child feel a bit odd in one way or another about using a language other than English.

In terms of your experience as a parent, we often call this the "grocery store factor." Many parents feel reasonably confident and okay using a language other than English with their child in private; however, in public—for example, in the grocery store checkout line—they can become much more self-conscious. As Emily, mother of two-year-old Ben, explained, "I just feel like I'm sort of being 'hoity-toity' or something—I mean here I am, this blond, Anglo woman, using Spanish in Safeway with my kid. As it's clearly not my first language, I just sometimes feel a little show-off-y or something."

Using a language other than English can make you stand out a bit, not just at the checkout line, but also in social situations like children's birthday parties. As Graham got to be a toddler, for instance, it became clearer to Kendall that much of her talk to Graham on such occasions was *meant* to be heard by other kids and mothers. She found herself switching to English to say things like "Okay, we're all sharing the trains" or "Wow, what a long train track you've all made." For Kendall, sticking only to Spanish in such contexts felt like it would keep them from being cooperative members of the group, since no one else would understand anything. It can also be difficult for non-native speakers of a language to speak to their children in front of native speakers if they don't feel 100 percent confident about their language skills.

Children likewise experience pressure to fit in, and such pressure typically increases as children grow, and as they spend more and more time with peers. While there are no quick fixes here, as a parent you do have the opportunity to frame this difference in positive terms for your child. Most parents find it helpful to explicitly acknowledge and discuss the ways in which they are special and different from other families, stressing the positive aspects of this difference (but also that they are not unique in their uniqueness—each family has its special characteristics). Some reverse psychology can also work here: one mother explained to us how whenever her school-age children respond to her Spanish by using English, she says, "Okay, if you want to be monolingual and be just like everyone else, we can speak English." (Her children soon realized themselves that they'd be losing out in the deal!)

If it's you and not your child who is having a hard time, remind yourself of the advantages for your child that come with bilingualism. And by now in this book you know that your practices are well supported by lots of research findings and the results of many other families. Considering all the important benefits that come with bilingualism, the occasional raised eyebrow at the grocery store is certainly a small price to pay for a lifetime of advantages for your child!

Another way of making this path a bit smoother is to simply explain what you are doing and why. In our experience, most people are merely curious and interested (and a few are impressed). Also, keep in mind that if you do find yourself moving back and forth in some situations (like the birthday party example), it's not a disaster or total failure. Your child learns some important lessons here as well, such as how different languages are used in different contexts with different people; how using the language that others know best is most polite; and how it's possible to move back and forth between two languages.

WHAT IF YOUR FAMILY CIRCUMSTANCES CHANGE?

We all experience major upheavals in our lives at some point or another. For example, according to the U.S. Department of Labor Bureau of Labor Statistics, 3.8 million people lost their jobs between 2003 and 2005 (that's about 1.6 million per year); many more people opt to leave their jobs on their own for something better. Lost jobs can mean lost resources and less money for things like a Spanish for toddlers program.

Other big changes include moves. One in six Americans move each year, and Americans on average move twelve times over the course of a lifetime. Families who move from the city to suburbs might find that they've completely lost their support system for Chinese, as there are no classes, no playgroups, and no Chinese-speaking babysitters in their area. The birth of a second or third child with the accompanying drain on time may also be a problem. Sometimes illness intervenes, either of another child or an aging parent, and suddenly, second language learning seems like a luxury and one which can be foregone.

Divorce is also common and is something that may also impact your efforts to raise a bilingual child. For example, a divorced U.S. mother who hoped her child would learn French from her husband and husband's family might well wonder if that plan went right out the window (along with her marriage) when her husband returned to Paris. Another mother we know named Gwen recently went through a bitter divorce. Her ex, Ferenc, a native speaker of Hungarian, has returned to Hungary, and Gwen has completely severed ties with both him and his family. However, Gwen and Ferenc have a five-year-old son, Laszlo, whom they had been raising bilingually in English and Hungarian. Gwen now lives with Laszlo in a small town in Oklahoma. She does not know of any speakers of Hungarian within a hundred-mile radius, the local library doesn't have any books or materials on Hungarian, and she doesn't have the financial resources to either find a tutor or buy expensive books online. In addition, she is not all that keen anymore on the Hungarian language in general and is wondering if it is really worth all the time and effort to keep Laszlo bilingual. "Wouldn't it just be easier to switch to English only?" she asks. "I mean, when is he ever going to use Hungarian *here*?"

Gwen is asking some important questions. Ambivalent feelings about bilingualism, lack of resources, lack of support—these are very real concerns for many people. But at the same time, they need not be insurmountable roadblocks either. Where there is a will, there is a way, and even when the will is somewhat lacking, there is still a way. One thing Gwen needs to realize is that, as we keep saying, there are many advantages to knowing more than one language: cognitive advantages, benefits in terms of cultural understanding, and, for Laszlo, the possibility of connecting with an important part of his heritage in the future.

If a parent has ambivalent feelings about a particular language (due to it being associated with a now decidedly less-than-beloved ex-spouse), the parent may still wish to consider the child's point of view. Even though the parent may not wish to see the ex again, the child may wish to keep in contact with the parent who does not have custody. And in some cases, due to court rulings, periodic contact may be necessary. Maintaining the child's ability to communicate in the other parent's language may be the best option.

Even when things are not going smoothly (due to moves, lost jobs, sick parents, or the birth of another child), parents may wish to consider whether bilingualism is really something they have to give up. You don't need expensive resources to raise a bilingual child. The most important thing is talking to your child yourself, and getting input for your child as best you can. Here are some other tips and suggestions that other parents in similar situations have shared with us over the years.

First, dial back the anxiety and remember that language learning is a lifelong process. A short break is not a total disaster. You and your children may need a little time to make changes and adjust, but new routines can be good.

Second, keep in mind that you and your child can study the language together, using resources from the Internet or the local library, or things that you have made at home (for example, simple board games). It may not be much for a while, but you can always keep on the lookout for new opportunities. In addition, as with the first point, don't stress. Whatever you can do is helpful, and you shouldn't fret over not doing enough.

Third, try to think about a change as a chance to shake up your language learning environment and support. Chances are that you were going to need to revise and update your plan as soon as your child got older (she couldn't stay in that Korean-for-toddlers class forever anyway!).

Fourth, look for new resources, new playgroups, and new classes. You may find yourself in a more supportive environment than you were previously. Finding new language learning contacts for your child may also help you as a parent adjust to your new home or lifestyle. Even if you are in a place that seems to be completely monolingual English, keep looking. Put up signs in grocery stores, at the local library, and at schools. Chances are, there are speakers out there, or other people interested in the same language.

Fifth, if you are divorced, negotiate and forge new relationships with your ex's family if possible. You'll need to establish your own lines of communication. Most grandparents are anxious about what will happen to their relationship with their grandchild after divorce. Take the initiative and approach them, and be explicit about your language learning hopes.

Sixth, if appropriate, consider switching languages. If you've moved from New York City to Quebec, for example, it may be a terrific opportunity for your child to switch from learning Chinese to learning French. The early exposure to Chinese certainly will not hurt your child, and the new environment may be optimal for acquiring French.

Lastly, check out different ways of practicing the language. More and more (free) language learning materials are on the Internet these days, and monthly Web access (with some companies) costs less than a few gallons of gas. Most local libraries, even in the smallest of towns, have computers available for public use. In addition, you can always request that your local library borrow material (such as books in the language you are interested in) from other libraries (an interlibrary loan).

Let's take Robert as an example. Robert and his ex-wife had adopted a two-year-old girl, Oksana, from Russia several years ago. Robert now has custody of Oksana, and though he is not a native speaker of Russian and in fact knows little Russian at all, he has made a concerted effort to help Oksana learn both Russian and English. Robert made friends with Russians in the area and took Oksana to local cultural events at the Russian embassy and Russian churches. After two years in the United States, Oksana was proud that she could speak Russian as well as English. Robert attributed this success to finding strong Russian-speaking role models in the community for his daughter and keeping her involved in the Russian community in the region.

WRAP UP

We've never met a parent who is raising a bilingual child who hasn't met a few bumps in the road. We hope this chapter has provided some strategies for navigating this path and the determination to forge ahead!

 POINTS TO REMEMBER:

- Try not to let language become a battleground for other issues in your family.
- You can't make your children speak a certain language. You can set the stage so that they want to speak it!
- Remain flexible and creative: Try out and experiment with the different tips and suggestions offered by parents in this chapter. They'll work differently for different kids at different ages.

CONCLUSION

The Bilingual Edge
Is for Everyone

We hope this book has made clear that bilingualism is not just desirable but also *possible*, and that the many benefits of bilingualism are backed by the latest research from linguistics, education, psychology, and cognitive science. We live in a rapidly changing world where international travel is commonplace, where business regularly crosses national borders, and where art (including cinema, music, literature) knows few boundaries. Being bilingual or even trilingual gives an individual the ability not only to understand other cultures, but also to participate more actively in these international endeavors. Knowing a second language has personal and familial benefits as well. It enhances creativity and academic success, it makes connections between generations stronger, and it allows an individual to connect with more of the world. It is a "boundary eraser" in all senses of the word.

As you know now, second language learning—like just about every type of learning—is most effective when it is enjoyable for everyone involved. Second language learning can be a fun and exciting family activity. It is most effective and enjoyable when it's integrated into everyday routines and interactions and when it's meaningfully connected with real life. These important points—well supported by a vast array of research—will help you successfully introduce a second language to your child.

MAKE LANGUAGE LEARNING
ENJOYABLE FOR EVERYONE INVOLVED

There are many fun ways parents can incorporate the second language into daily family routines to help establish positive associations with the language from a young age. For instance, all parents can read simple books with their children. This is an enjoyable and relaxing way not just to bond with your young child, but also to boost second language vocabulary. As children get a bit older, you can incorporate different types of edutainment. Used carefully, movies, videos, and computer games can help encourage your child to interact with you and others in the second language and also serve as great motivators.

Fun and enjoyable should also be the guiding criteria when you are deciding on educational programs. A huge number of resources exist for lots of different languages. These range from heritage schools, immersion camps, and courses at cultural centers, to bilingual education programs in public and private schools to tutors and language classes for babies and children. Whatever you are looking for, choose a program that has clearly defined goals, emphasizes bilingualism, biliteracy, and biculturalism, has enthusiastic teachers, and actively *uses* the target language (as opposed to only formally teaching it). Take time to find the program that is right for you and your family by observing a lesson, asking about the curricula, seeing if the program is in sync with your beliefs about language learning and if you think it will fit in with your child's needs and personality.

INTEGRATE LANGUAGE LEARNING INTO
EVERYDAY ROUTINES AND INTERACTIONS

It's a relief for most busy parents to find out that second language learning doesn't always have to take huge amounts of extra time. Modern life and parenting are hectic and stressful enough as it is! Second language learning works best when it is integrated into everyday routines and interactions. For young children, all language learning happens through the course of everyday conversations with caretakers, for instance, as they are being changed, being fed, or being pushed around the grocery store.

The most important thing parents can do to promote language learning is to talk to their child as much as possible in the languages they want them to learn. Both quantity and quality of talk are important here!

MAKE LANGUAGE LEARNING MEANINGFUL
AND CONNECTED WITH REAL LIFE

Language learning works best when it's situated in the real world and through real interactions. Family and community can help connect language learning to real life (and real needs to use the language). Having other speakers of that language nearby can provide a huge incentive for both you and your child, and also means that there are possibilities for bilingual playgroups, reading hours, story time, babysitters, and playdates. For a young child, what could be more meaningful than using that language to communicate with a new friend?

By the same token, if you have a family connection to a particular language, this is also an important factor. Being able to use the language with grandparents or great aunts, for instance, connects the language in real and emotional ways with everyday life.

LANGUAGE LEARNING IS A LIFELONG PROCESS

Language learning is a lifelong process in many respects. How well your children speak their second or third languages will vary depending on their environment and its interactional demands. There's a tendency for all of us, however, to think of bilingualism as a fixed goal or milestone. Second language learning differs from other milestones such as learning to walk or getting dressed on one's own—it's a lifelong activity. The downside here is that it's never "done." If bilingualism is a goal, parents always need to keep an eye on both the quantity and quality of language exposure for their children. The plus side here is that it is never too late!

This long-term perspective is important because there will be good days and bad days. Your child may refuse to speak her second language for weeks, months, or even years. Get ready for this (it's pretty much inevitable) and be well prepared, but don't take it to heart! This doesn't mean she doesn't understand the language, and it certainly doesn't mean she's not benefiting from exposure to it. There are also big differences between how children progress—even under very similar language learning environments! Many things can influence a child's language development—birth order, gender, aptitude, personality, parenting style, and so on—but the differences are usually minor and even out over time.

Remember, even if your child goes on to study a totally new and different third language later in life, the investment in her second language is hardly wasted. Indeed, there is evidence that knowledge of a second language promotes acquisition of a third. Early second language learning is a good investment as second language learning strategies seem to transfer across languages.

Every child is unique and every family situation is different. Which language you choose, how you introduce and balance different languages in your home, and how your child progresses are personal, individual decisions and stories. We hope this book has helped you to see how to make the most of *The Bilingual Edge* for *your own* child. Our very best wishes as you continue on this unique and exciting journey!

ACKNOWLEDGMENTS

The insight and inspiration to write this book was born with Miranda Mackey Yarowsky and Graham King Heathcote. Our first debt of gratitude is to them, to their fathers, and to our own mothers.

This book has benefited from outstanding research support. Rebekha Abbuhl, Lyn Fogle, and Rebecca Sachs helped in critical ways. We also gratefully acknowledge the important work of Moonjeung Chang, Amy Firestone, Jaemyung Goo, Mika Hama, Jamie Lepore Wright, Aubrey Logan-Terry, Deborah Petit, Jamie Schissel, Abbe Spokane, and Boram Suh.

We are lucky to have many friends and colleagues whose interests overlap with ours and who read drafts of chapters in this book, many of whom provided excellent comments and suggestions. We thank Hilary Appel, Hidy Basta, Carol Benson, Cathrine Berg, Heidi Byrnes, Helen Carpenter, Luis Cerezo, Debby Chen Pichler, Lena Edlund, Miriam Eisenstein Ebsworth, Shiraz Felling, Akiko Fujii,

Susan M. Gass, Breda Griffith, Jennifer Leeman, Gigliana Melzi, Yoshiko Mori, Ana-Maria Nuevo, Marisol Perez Casas, Rhonda Oliver, Jenefer Philp, Becky Resnik, Lauren Ross-Feldman, Maria Sáez-Martí, Nuria Sagarra, Cristina Sanz, Rita Silver, Maria Snyder, Deborah Tannen, and Paula Winke.

Finally, we thank our agents, Stefanie Von Borstel and Lilly Ghahremani of Full Circle Literary, and our editor, Laura Dozier of HarperCollins, for believing in the importance of this project and working enthusiastically with us every step of the way.

REFERENCES AND RESOURCES

*These are the essential references and resources
discussed in the book. Please also see
our Web site—www.thebilingualedge.com—for
more detailed information and further
references and resources.*

GREAT WEB SITES FOR FOREIGN LANGUAGE BOOKS

Amazon.com's Foreign Books Section
http://www.amazon.com/exec/obidos/tg/browse/-/3118571/102–
9852915–3125710
This page on the Amazon site allows the user to browse books by
language. There are twelve languages listed on this page, plus a
link to more languages, including more than sixty others. The site
can also perform individualized searches, so that a search for
"children's Spanish books" yields Spanish-language textbooks,
dual language story books, Spanish language versions of English
language books like *Goodnight Moon*, and more.

Books Without Borders
http://www.bookswithoutborders.com/
This site, specifically geared toward parents raising bilingual or
multilingual children, offers a wide selection of books in English,
Spanish, Russian, German, French, Italian, and Chinese. This site
also offers foreign language video- and audiocassettes.

Children's Foreign Language Bookstore (in association with Amazon.com)

http://www.geocities.com/Athens/Delphi/1794/flbooks.html

This site provides foreign language books, songs, and games for parents. The resources on this site are primarily in Spanish, French, German, and American Sign Language.

Language Book Centre

http://www.languagebooks.com.au/

This Australian site has foreign language titles for children and adults as well as foreign language textbooks, videos, and DVDs.

Powell's Books—Foreign Languages

http://www.powells.com/psection/ForeignLanguages.html

This page specializes in foreign language books. Links on the side of the page can refer you to books in various languages and genres.

Reforma

http://www.reforma.org

An affiliate of the American Library Association that provides resources for Spanish-English bilingual children and their parents.

Schoenhof's Foreign Books

http://www.schoenhofs.com/

This site offers books in more than forty languages. It provides searches by individual language in the categories of fiction and nonfiction, children's books, and texts for language learning.

PARENT GUIDELINES CONCERNING THE IMPACT
OF THE MEDIA ON CHILD DEVELOPMENT

Parent Center
http://parentcenter.babycenter.com/refcap/64211.html

Clearinghouse on Early Education and Parenting
http://ceep.crc.uiuc.edu/eecearchive/digests/1990/famtv90.html

Good News Family Care (U.K.)
http://gnfc.org.uk/tv_manag.html

Educational Research Information Clearinghouse
http://www.ericdigests.org/pre-9215/family.htm

Child Development Institute
http://www.childdevelopmentinfo.com/health_safety/television
.shtml

Entertainment Software Rating Board (ESRB)
http://www.esrb.org/
http://www.eslcafe.com/forums/student/

RESOURCES FOR FOREIGN LANGUAGE DVDS

Region Code Free DVD Players (http:// www.regioncodefreedvd.com)
and Code Free DVD (http://www.codefreedvd.com)
These Web sites specialize in selling products that allow you to
watch any DVD on any TV.

HackDVD (http://www.videohelp.com/dvdhacks)
This Web site offers instructions on how to make existing DVD
players region free, often by entering a simple code on the remote.

262   •   REFERENCES AND RESOURCES

Note: Doing this usually invalidates your warranty *and may harm your DVD player in some cases.*

DVDs on Your Computer

There are also a number of software applications that allow you to play any type of DVD on your computer. One such application for PCs is AnyDVD, which can be purchased at http://www.slysoft .com/.

Jump TV (http://www.jumptv.com)

This Web site offers samples of TV programs from around the world.

RESOURCES FOR HERITAGE SCHOOLS

- United States database for heritage schools: http://www.cal.org/ heritage/programs/profiles/index.html
- French Heritage Language Program/French-American Cultural Exchange (FACE) in New York City: http://www.cal.org/heritage/ programs/profiles/FrenchAmCultEx.html
- Spanish Literacy Squared (San Diego, California): http://www.cal .org/heritage/programs/profiles/LiteracySquared.html
- Mandarin Chinese heritage program (Chong Wa Education Society, Seattle, Washington): http://www.cal.org/heritage/ programs/profiles/ChongWa.html
- Chinese heritage schools in Northern California: http://www .anccs.org/05/2005member.htm
- Japanese heritage schools in the United States (with a clickable map): http://www.colorado.edu/ealld/atj/SIG/heritage/webmap/ map.html
- Heritage language schools in Canada: http://www.settlement .org/sys/link_redirect.asp?doc_id=1001593
- Chinese School Association in the United States: http://www .csaus.org—Mainland Chinese, simplified characters; can search for schools by state. Some listings include a link to the Web site for the specific school.

RESOURCES FOR HERITAGE AND IMMERSION CAMPS

- Examples of heritage camps for adoptive families: http://www.heritagecamps.org/camps.html
- Concordia language villages: http://clvweb.cord.edu/prweb/

RESOURCES FOR PARENTS OF CHILDREN CONCERNED WITH FIRST LANGUAGE DEVELOPMENT

- *Childhood Speech, Language & Listening Problems: What Every Parent Should Know* by Patricia McAleer Hamaguchi. 1995. John Wiley & Sons Inc.
- *Beyond Baby Talk: From Sounds to Sentences, A Parent's Complete Guide to Language Development* by Kenn Apel, Ph.D. and Julie Masterson, Ph.D. 2001. Three Rivers Press.

BOOKS FOR PARENTS OF CHILDREN WITH SPECIAL NEEDS

- *The Child with Special Needs: Encouraging Intellectual and Emotional Growth* by Stanley I. Greenspan, Serena Wieder, and Robin Simons. 1998. Perseus Books.
- *Simple Successes: From Obstacles to Solutions with Special Needs Children* by Rachelle Zola. 2006. Outskirts Press.
- *Preschool Children with Special Needs: Children at Risk, Children with Disabilities* (2nd Edition) by Janet W. Lerner, Barbara Lowenthal, and Rosemary W. Egan. 2002. Allyn & Bacon.
- *Married with Special-Needs Children: A Couples' Guide to Keeping Connected* by Laura E. Marshak and Fran P. Prezant. 2007. Woodbine House.
- *Young Children with Special Needs* by Richard Gargiulo and Jennifer L. Kilgo. 2004. Thomson Delmar Learning.
- *Breakthrough Parenting for Children with Special Needs: Raising the Bar of Expectations* by Judy Winter. 2006. Jossey-Bass Publishers.

EXERCISE WORKSHEETS AND SAMPLE ANSWERS

CHAPTER 1

EXERCISE

Why Do You Want Your Child to Speak More Than One Language?

1. I want my child to become bilingual because it will enhance creativity.

Strongly agree   Somewhat agree   Somewhat disagree   Strongly disagree

Why?_____

2. I want my child to know more than one language so she can understand more about other cultures.

Strongly agree   Somewhat agree   Somewhat disagree   Strongly disagree

Why?_____

3. I want my child to know my heritage language so she can participate fully in conversations with our extended family.

Strongly agree   Somewhat agree   Somewhat disagree   Strongly disagree

Why?_____

4. I hope that bilingualism might help my child learn to read more easily.

Strongly agree   Somewhat agree   Somewhat disagree   Strongly disagree

Why?_____

5. I want my child to know something of his heritage, and knowing his heritage language is an essential part of that.

Strongly agree   Somewhat agree   Somewhat disagree   Strongly disagree

Why?_____

6. Bilingualism is essential for living in a multicultural, multilingual world.

Strongly agree   Somewhat agree   Somewhat disagree   Strongly disagree

Why?_____

7. I want my child to know more than one language so that he can make friends across cultural lines.

Strongly agree   Somewhat agree   Somewhat disagree   Strongly disagree

Why?_____

8. I think bilingualism is important for my child because it will help her academically.

Strongly agree   Somewhat agree   Somewhat disagree   Strongly disagree

Why?_____

9. I want my child to know my heritage language because it is the language of intimacy and affection in my heart and mind.

Strongly agree   Somewhat agree   Somewhat disagree   Strongly disagree

Why?_____

KEY: If you strongly agree or somewhat agree with statements 1, 4, 8: You are very interested in the academic and cognitive advantages associated with knowing more than one language. In order to get the maximum edge, a child must achieve relatively high levels of competence in that language. This book is written with that aim in mind and to help you meet that goal.

If you strongly agree or somewhat agree with statements 2, 6, 7: You are deeply concerned with the cultural, interrelational, and communicative advantages that come with knowing two or more languages. Your family is interested in other cultures and committed to breaking down barriers between groups. We take a similar perspective in this book and share like-minded goals.

If you strongly agree or somewhat agree with statements 3, 5, 9: Your child's cultural heritage is important to you, and you see language as an important piece of that heritage. You and your family recognize that language learning is intimately and personally connected to identity. Each family is unique. *The Bilingual Edge* shows families how to utilize their particular family qualities and characteristics to maximize language learning potential.

CHAPTER 4 EXERCISE SAMPLE ANSWERS
FROM PAGE 76

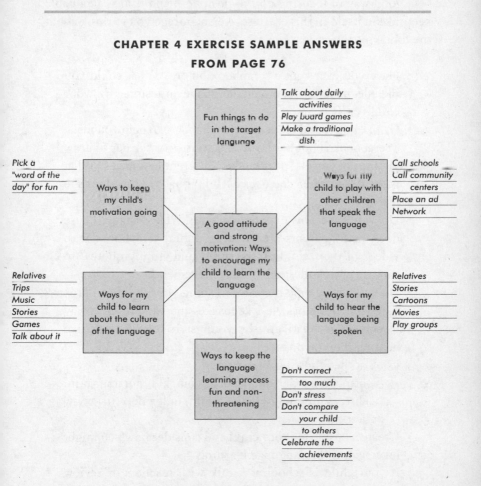

CHAPTER 5 FOLLOW-UP EXERCISE:
LEVERAGING INDIVIDUAL DIFFERENCES
FOR LANGUAGE LEARNING

We have emphasized the importance of refraining from comparisons and celebrating your child's achievements. Having a record of your child's achievements (both in language and in other areas) can be a good way of focusing on the positive—and is also a valuable keepsake in itself. In this exercise, we encourage you to think about the following questions.

- Have you ever kept a scrapbook before? What could you include in it to document language milestones in your child's early years?
- Or, ask your child (depending on her age) to help you make a scrapbook about her favorite activities or ask her to draw pictures of herself speaking her different languages (What is she doing when she uses each language? How does she feel?).
- Do you have any experience with Web design? These days, many parents are creating digital scrapbooks with videos of their children, pictures, and "funny quotes" to capture early milestones and achievements.
- Do you have a tape recorder? Taping your child can provide you with a priceless keepsake, and can also be a way to encourage relatively shy youngsters to talk (kids like hearing their own voices).
- Look at the following Web sites:
    (i) http://www.personalitypage.com/kid_portraits.html
    (ii) http://uniquelyyou.net/child/index.php (iii) www
    .milest.com
- Create a profile for your child and consider how you might leverage this for language learning.
- If your child is school-age, talk with teachers or review reports about your child's learning preferences. What

strengths have been identified and how do these coincide with his or her language learning based on what you know now?

- Some researchers have referred to language learners as auditory, visual, kinesthetic, and tactile. Think about your child from this perspective. Refer to the diagram earlier in this chapter for help.
- Think about what you can learn about your child from his or her language use. Note what types of expressions or phrases he or she uses frequently and consider the reasons why. What new activities, games, or learning environments may help expand his or her abilities?

CHAPTER 9 EXERCISE:
A MINI-RESEARCH PROJECT—ASSESSING
FAMILY LANGUAGE USE
FROM PAGES 199 AND 120

FAMILY LANGUAGE USE			
Situation:		Date and Time:	
Part 1			
Mother:			
Father:			
	Language 1	**Language 2**	**Total**
Number of words (%)			
Words by Father (%)			
Words by Mother (%)			
Frequently used words (and phrases if applicable)			
Part 2			
	Child	**Mother**	**Father**
Number of code-switches			
L1 to L2			
L2 to L1			
Reaction to Switch			

REFERENCES

CHAPTER 1

Bialystok, E. (2001). *Bilingualism in Development*. Cambridge: Cambridge University Press.

Bruck, M., & Genesee, F. (1995). Phonological awareness in young second language learners. *Journal of Child Language, 22*, 307–324.

Campbell, R., & Sals, E. (1995). Accelerated metalinguistic (phonological) awareness in bilingual children. *British Journal of Development Psychology, 13*, 61–68.

Cazabon, M., Lambert, W., & Hall, G. (1992). *Two-way Bilingual Education: A Progress Report on the Amigos Program*. Santa Cruz, CA and Washington, DC: National Center for Research on Cultural Diversity and Second Language Learning.

Cummins, J. (1978a). Bilingualism and the development of metalinguistic awareness. *Journal of Cross-Cultural Psychology, 9*, 131–149.

Eviatar, Z., & Ibraham, R. (2000). Bilingual is as bilingual does: Metalinguistic abilities of Arabic-speaking children. *Applied Psycholinguistics, 21*, 451–471.

Galambos, S. J., & Goldin-Meadow, S. (1990). The effects of learning two languages on metalinguistic awareness. *Cognition, 34*, 1–56.

Gopnik, A., & Choi, S. (1990). Do linguistic differences lead to cognitive differences? A cross-linguistic study of semantic and cognitive development. *First Language, 10,* 199–215.

Ianco-Worrall, A. (1972). Bilingualism and cognitive development. *Child Development, 43,* 1390–1400.

Ogbu (1987). Variability in minority responses to schooling. A problem in search of an explanation. *Anthropology & Education Quarterly, 18,* 312–334.

Peal, E., & Lambert, W. E. (1962). The relation of bilingualism to intelligence. *Psychological Monographs, 76*(27), 1–23.

Tawada, Y. (2003). Writing in the web of words. In I. de Courtivron (Ed.), *Lives in Translation: Bilingual Writers on Identity and Creativity* (pp. 147–155). New York: Palgrave Macmillan.

U.S. Census Bureau (2005). *American Community Survey.* Retrieved on October 12, 2006, from http://www.census.gov/acs/www/.

CHAPTER 2

Slobin, D. I. (1994). *Crosslinguistic Aspects of Child Language Acquisition (Sophia Linguistica Working Papers in Linguistics No. 35).* Tokyo: Sophia University.

CHAPTER 3

Johnston, J. C., Durieux-Smith, A., & Bloom, K. (2005). Teaching gestural signs to infants to advance child development: A review of the evidence. *First Language, 25 (2),* 235–251.

Kenner, C., & Kress, G. (2003). The multisemiotic resources of biliterate children. *Journal of Early Childhood Literacy, 3(2),* 179–202.

CHAPTER 4

Ioup, G., Boustagui, E., Tigi, M., & Moselle, M. (1994). Reexamining the critical period hypothesis: A case study in a naturalistic environment. *Studies in Second Language Acquisition, 16,* 73–98.

Pearson, B. Z., Fernandez, M. C., & Oller, D. K. (1993). Lexical development in bilingual infants and toddlers: Comparison to monolingual norms. *Language Learning, 43,* 93–120.

Saxton, M., Backley, P., & Gallaway, C. (2005). Negative input for grammatical errors: Effects after a lag of 12 weeks. *Journal of Child Language, 32(3),* 643–672.

Snow, C., & Hoefnagel-Höhle, M. (1977). Age differences in the pronunciation of foreign sounds. *Language and Speech, 20,* 357–365.

CHAPTER 5

Adams, A M., & Gathercole, S. E. (1996). Phonological working memory and spoken language development in young children. *The Quarterly Journal of Experimental Psychology, 49A(1),* 216–233.

Bauer, D. J., Goldfield, B. A., & Reznick, J. S. (2002). Alternative approaches to analyzing individual differences in the rate of early vocabulary development. *Applied Psycholinguistics, 23,* 313–335.

Bongartz, C., & Schneider, M. L. (2003). Linguistic development in social contexts: A study of two brothers learning German. *The Modern Language Journal, 87(1),* 13–37.

DeKeyser, R. M. (2000). The robustness of critical period effects in second language acquisition. *Studies in Second Language Acquisition, 22(4),* 499–533.

Felling, S. (2006). *Fading Farsi: Language Policy, Ideology, and Shift in the Iranian American Family.* Unpublished doctoral dissertation, Georgetown University.

Geschwind, N., & Galaburda, A. M. (1985). Cerebral lateralization: Biological mechanisms, associations, and pathology. A hypothesis and a program for research. *Archives of Neurology, 42,* 428–459.

Goldfield, B. A., & Snow, C. E. (2004). Individual differences in language development. In J. Berko Gleason (Ed.), *The Development of Language, 6th edition* (pp. 315–346). Boston: Allyn & Bacon.

Goodwin, M. H. (2001). Organizing participation in cross-sex jump rope: Situating gender differences within longitudinal studies of activities. *Research on Language and Social Interaction, 34(1),* 75–106.

Harley, B., & Hart, D. (1997). Language aptitude and second language proficiency in classroom learners of different starting ages. *Studies in Second Language Acquisition, 19,* 379–400.

Javorinskij, A. (1979). On the lexical competence of bilingual children of kindergarten-age groups. *International Journal of Psycholinguistics, 6(3)*, 43–57.

Karrass, J., & Braungart-Rieker, J. M. (2003). Parenting and temperament as interacting agents in early language development. *Parenting, 3(3)*, 235–259.

Kyratzis, A. (2000). Tactical uses of narratives in nursery school same-sex groups. *Discourse Processes, 29(3)*, 269–299.

Nelson, K. (1973). Structure and strategy in learning to talk. *Monographs of the Society for Research in Child Development, 38(1/2)*, 1–135.

Pine, J. M. (1995). Variation in vocabulary development as a function of birth order. *Child Development, 66*, 272–281.

Ramírez-Esparza, N., Gosling, S. D., Benet-Martínez, V., Potter, J., & Pennebaker, J. W. (2006). Do bilinguals have two personalities? A special case of cultural frame switching. *Journal of Research in Personality, 40*, 99–120.

Schachter, F. F., Shore, E., Hodapp, R., Chalfin, S., & Bundy, C. (1978). Do girls talk earlier? Mean length of utterance in toddlers. *Developmental Psychology, 14*, 388–392.

Shin, S. J. (2002). Birth order and the language experience of bilingual children. *TESOL Quarterly, 36(1)*, 103–113.

Skehan, P. (1986). The role of foreign language aptitude in a model of school learning. *Language Testing, 3(2)*, 188–221.

Ullman, M. T., Estabrooke, I. V., Steinhauer, K., Brovetto, C., Pancheva, R., Ozawa, K., Mordecai, K., & Maki, P. M. (2002). Sex differences in the neurocognition of language. *Brain and Language, 83*, 9–224.

Wong-Fillmore, L. (1976). *The Second Time Around: Cognitive and Social Strategies in Second Language Acquisition.* Unpublished doctoral dissertation, Stanford University.

CHAPTER 6

Berko-Gleason, J. (1975). Fathers and other strangers: Men's speech to young children. In D. P. Dato (Ed.), *Georgetown University Roundtable* (pp. 280–289). Washington, D.C.: Georgetown University.

Dolson, D. P. (1985). The effects of Spanish home language use on the scholastic performance of Hispanic pupils. *Journal of Multilingual and Multicultural Development*, *6(2)*, 135–155.

High, P. C., LaGasse, L., Becker, S., Ahlgren, I., & Gardner, A. (2000). Literacy promotion in primary care pediatrics: Can we make a difference? *Pediatrics*, *105(4)*, 927–934.

Moore, S. (November, 2004). "Grandcare: Encouraging that special relationship between your parents and your kids." *Washington Parent*. Retrieved on October 16, 2006, from http://www.washingtonparent.com/articles/0411/pep.html.

Patterson, J. L. (1998). Expressive vocabulary development and word combinations of Spanish-English bilingual toddlers. *American Journal of Speech-Language Pathology*, 7, 46–56.

Pearson, B. Z., Fernandez, S. C., Lewedeg, V., & Oller, D. K. (1997). The relation of input factors to lexical learning by bilingual infants. *Applied Psycholinguistics*, *18*, 41–58.

Rumbaut, R. G., Massey, D. S., & Bean, F. D. (2006). Linguistic life expectancies: Immigrant language retention in Southern California. *Population and Development Review*, 32 (3), 447–460.

Skutnabb-Kangas, T., & Toukomaa, P. (1976). Teaching migrant children's mother tongue and learning the language of the host country in the context of the sociocultural situation of the migrant family. Helsinki: The Finnish National Commission for UNESCO.

Snow, C. (1972). Mothers' speech to children learning languages. *Child Development*, *43*, 549–565.

CHAPTER 7

American Academy of Pediatrics (2006). "Television and the family." Retrieved October 22, 2006, from http://www.aap.org/family/tv1.htm.

Eliot, L. S. (1999). *What's Going on in There? How the Brain and Mind Develop in the First Five Years of Life*. New York: Bantam.

Kuhl, P. K. (2004). Early language acquisition: Cracking the speech code. *Nature Reviews Neuroscience*, *5*, 831–843.

Patterson, J. L. (1998). Expressive vocabulary development and word combinations of Spanish-English bilingual toddlers. *American Journal of Speech-Language Pathology*, 7, 46–56.

CHAPTER 8

Center for Applied Linguistics (2006). *Directory of two-way bilingual immersion programs in the U.S.* Retrieved October 13, 2006, from http://www.cal.org/twi/directory.

Liu, P. (2006). Community-based Chinese Schools in Southern California: A Survey of Teachers. *Language, Culture, and Curriculum, 19*, 237–247.

National Clearinghouse for English Language Acquisition and Language Instruction Educational Programs (2006). Glossary of terms related to the education of linguistically and culturally diverse students. Retrieved October 22, 2006, from http://www.ncela.gwu.edu/expert/glossary.html.

Rolstad, K., Mahoney, K., & Glass, G. V. (2005). The big picture: A meta-analysis of program effectiveness research on English language learners. *Educational Policy, 19*, 572–594.

Rosenbusch, M. (1995). Guidelines for starting an elementary school foreign language program. Washington, D.C.: Center for Applied Linguistics. Retrieved October 15, 2006, from http://www.cal.org/resources/digest/rosenb01.html.

Senesac, B. (2002). Two-way bilingual immersion: A portrait of quality schooling. *Bilingual Research Journal, 26(1)*, 1–17. Retrieved from http://brj.asu.edu/content/vol26_no1/html/art6.htm.

Swain, M. (1985). Communicative competence: Some roles of comprehensible input in its development. In S. M. Gass, & C. G. Madden (Eds.), *Input in Second Language Acquisition* (pp. 235–53). Rowley, MA: Newbury House.

CHAPTER 9

Lanza, E. (1997). *Language Mixing in Infant Bilingualism: A Sociolinguistic Perspective.* New York: Oxford University Press.

Peal, E., & Lambert, W. E. (1962). The relation of bilingualism to intelligence. *Psychological Monographs, 76(27)*, 1–23.

Poplack, S. (1982). Sometimes I'll start a sentence in Spanish y termino en español: Toward a typology of code-switching. In J. Amastae, & L. Elías-Olivares (Eds.), *Spanish in the United States: Sociolinguistic Aspects* (pp. 230–263). Cambridge: Cambridge University Press.

Zentella, A. C. (1997). *Growing up Bilingual: Puerto Rican Children in New York*. Malden, MA: Blackwell.

CHAPTER 10

Apel, K., & Masterson, J. (2001). *Beyond Baby Talk: From Sounds to Sentences, a Parent's Complete Guide to Language Development*. New York: Three Rivers Press.

Gargiulo, R., & Kilgo, J. L. (2004). *Young children with special needs*. Florence, KY: Thomson Delmar Learning.

Greenspan, S. I., Wieder, S., & Simons, R. (1998). *The Child with Special Needs: Encouraging Intellectual and Emotional Growth*. New York: Perseus Books.

Hamaguchi, P. M. (2001). *Childhood Speech, Language & Listening Problems: What Every Parent Should Know*. Indianapolis: Wiley.

Krupa-Kwiatkowski, M. (1998). You shouldn't have brought me here!: Interaction strategies in the silent period of an inner-directed second language learner. *Research on Language and Social Interaction, 31(2),* 133–175.

Lerner, J. W., Lowenthal, B., & Egan, R. W. (2002). *Preschool Children with Special Needs: Children at Risk, Children with Disabilities* (2nd Edition). Boston: Allyn & Bacon.

Prezant, F. (2006). *Married with Special-Needs Children: A Couples' Guide to Keeping Connected*. Bethesda, MD: Woodbine House.

Toppelberg, C. O., Snow, C. E., & Tager-Flusberg, H. (1999). Severe developmental disorders and bilingualism. *American Academy of Child and Adolescent Psychiatry, 38(9),* 1197–1199.

U.S. Census Bureau (2005). *American Community Survey*. Retrieved on October 12, 2006, from http://www.census.gov/acs/www/.

U.S. Department of Labor Bureau of Labor Statistics (2006). *Bureau of Labor Statistics*. Retrieved on October 12, 2006, from http://www.bls.gov/.

Winter, J. (2006). *Breakthrough Parenting for Children with Special Needs: Raising the Bar of Expectation*. San Francisco: Jossey-Bass Publishers.

Zola, R. (2006). *Simple Successes: From Obstacles to Solutions with Special Needs Children*. Parker, CO: Outskirts Press.

REFERENCES

CHAPTER 11

Cenoz, J., & Valencia, J. (1994). Additive trilingualism: Evidence from the Basque Country. *Applied Psycholinguistics, 15*, 195–207.

Cruz-Ferreira, M. (2006). *Three Is a Crowd?: Acquiring Portuguese in a Trilingual Environment.* Clevedon, UK: Multilingual Matters.

Crystal, D. (1997). *The Cambridge Encyclopedia of Language.* Cambridge: Cambridge University Press.

DeWaele, J. M. (2000). Trilingual first language acquisition: Exploration of a linguistic "miracle." *LaChouette, 31*, 41–46.

Gordon, R. G., Jr. (Ed.) (2005). *Ethnologue: Languages of the World.* Retrieved from http://www.ethnologue.com/.

McCloskey, J. (2000). Quantifier float and wh-movement in an Irish English. *Linguistic Inquiry, 31(1)*, 57–84.

Quay, S. (2001). Managing linguistic boundaries in early trilingual development. In J. Cenoz, & F. Genesee (Eds.), *Trends in Bilingual Acquisition* (pp. 149–199). Amsterdam: John Benjamins.

INDEX

Academic
  ability viii
  achievement 115, 154–158
  grade point average 114
  growth 116
  skills 5, 69, 160
Accent x, 22, 55, 58, 61, 65–66, 71–73,
  102, 134, 238
Activities ix, 20, 80, 100, 132, 215, 269
  and babysitters 117, 120
  and family language audit 128–131
  and gender 85
  and grandparents 126
  and independent language teachers/
    tutors 173–177
  and personality 86
  and playgroups 121
  and language programs 170
  and learning style 91–94
  and television 138
  arts and crafts 117
  community 12
  cultural 53, 166
  everyday 104–105, 109
  interactive 29, 150, 267–268
  Internet-based 144–147

language games 213
language-based 43–48, 52, 101, 145
language-rich 81, 119, 123
rituals 107
routine 118
second language 99
Adoptive families 43, 165–166, 142,
  250–251
African American Vernacular English
  (AAVE) 232
Age (of child) 55–76, 84, 100, 240–242,
  244–246, 248, 253
  and accents 22, 55–58, 65–66, 71–72
  and aptitude 56, 65, 68, 90
  and edutainment 134–135, 144–150
  and language delay 208–209
  and more than two languages
    222–223, 226
  and social and emotional differences
    56–57, 73
  and special needs 216–218
  and speed of learning 21, 57–58, 68
  and television 136, 150
  and ultimate level of success 21–22
  eight-year-old children 73, 93,
    176–177, 191–192, 222, 237–238

Age (of child) (*continued*)
  eleven-year-old children 73, 146,
    237–238
  five-year-old children 6, 62–63, 166,
    207–208, 209, 239, 244, 248
  four-year-old children 38, 79–80,
    111, 128, 209, 226, 235, 241, 244–245
  infants 39–40, 51, 59–60, 83, 101,
    103, 136–137, 139
  one-year-old children 36, 59–60, 71
  school-age children 68–75, 143–150
  three-year-old children 62, 71, 78,
    80, 230
  toddlers 64, 103–104, 121, 131, 142,
    150, 233
  two-year-old children 78, 118–119,
    136, 187–188, 208, 214–215,
    222–223, 239, 243
Ahlgren, Ingrid 103–104
Allende, Isabelle 5
Alliance Francais 167
American Academy of Pediatrics
    135–136
American Sign Language (ASL) 52–54
Anxiety 183, 193
  children's 21, 212
  parents' 209, 249
Aptitude 56, 65, 68, 77, 84, 89–90, 255
Arabic language 47, 64–67, 156, 196,
    210, 224–225
Assessing language viii, 40–47, 71, 199
Assimilation 12, 169
Attention Deficit Hyperactivity
    Disorder 216, 236
Attention span 83
Attitudes 41, 44, 197, 103, 205
  positive 8, 15, 58, 69, 76, 160,
    174–175
Au pairs 116–119 *see also nannies*
Audit (family language) 107–110, 116,
    128–132, 233
Autism 209, 216
Azeri language 80

Babbling 27, 59, 189 *see also first language
    acquisition*
Baby Einstein 135 *see also videos and
    DVDs*

Baby sign language 49–54 *see also
    American Sign Language*
Baby videos 29, 150 *see also videos*
BabyFirst TV 29
Babysitters 44, 63, 100–104, 116–117,
    176–177, 190, 225, 248, 254 *see also
    nannies*
Babysitting coops 52
Balanced bilingual(ism) 112, 201–203,
    234
Banderas, Antonio 243
Basque language 228
Bauer, Daniel 83
Bean, Frank D. 112
Becker, Samuel 113
Benefits
  of bilingualism vii, 3–7, 15, 32, 38, 45,
    53, 114, 242, 247–249
Bialystok, Ellen 6–7
Bicultural(ism) 172, 253
Bilingual/bilingualism
  advantages of bilingualism *see benefits
    of bilingualism*
  bilingualism and academic skills 5,
    160
  bilingualism and advanced
    proficiency 3–4, 7, 15
  bilingual authors 5
  bilingualism and birth order 78–79,
    89, 255
  bilingual books 29, 259–260
  bilingual children 5–10, 18–20,
    26–27, 62, 107, 113, 127, 186–188,
    195, 201, 207–209
  bilingual cartoon characters *see also
    television*
    Diego 139
    Dora the Explorer 133
  bilingual dolls/toys vii, 20, 140–143
    Care Bears 141
    Dora the Explorer 20
    Language Littles 141
    Young Hee 141
  bilingual education 156–169, 228–229
  French-English bilinguals 240
  bilingual groups 123
  bilingual language development 27
    see also second language learning

bilinguals and literacy 3, 69, 142, 164, 168, 204
bilingualism and metalinguistic awareness 5–7, 229
bilinguals as compared to monolinguals 4–7, 12–15, 20, 26, 62, 115, 155–158, 188–189, 201–203, 207–210, 222, 228 *see also monolingual/monolingualism*
bilinguals and multiculturalism 8
bilingual parenting/parents 17–20, 107, 193
passive bilinguals 108–109, 129–132, 224–225
bilingual playgroups 88, 254
bilingual rebellion 240–241
Spanish-English bilinguals 112–113, 114, 190–192
bilingual speech specialists 209 *see also speech-language pathologists*
bilinguals and writing 5, 53, 171, 174
Bilingual education programs 18, 30–33, 52, 253
opposition to 154
types of 153–159
Biliteracy/biliterate 156–159, 168, 172, 253
Birth order 77–80, 89, 255
Blogs 145, 148
Brain development, early 136
Brainy Baby 135 *see also DVDs and videos*
British schools of America organization 164

Cabazon, Mary 9
California 30, 112, 154–158
Cambodian language 156
Canada 30, 50, 131, 167, 171, 204, 229, 243, 262
Canadian Association of Private Language Schools 167
CDs vii, 73, 147, 232 *see also DVDs*
Cenoz, Jasone 228
Center for Applied Linguistics (CAL) 161
Chinese language 24–25, 31, 38, 43, 47–49, 52–53, 131, 136,-137, 163, 221, 231

Choi, Soonja 10–11
Cisneros, Sandra 5
Code-mixing 185–189, 190–194, *see also code-switching*
Code-switching 190–206, 270 *see also code-mixing*
Cognitive flexibility 229
Commonsense beliefs about language learning 17
Communication strategies 149
Community viii, 9, 44–48, 100, 179, 185, 190–196, 218, 254
activities 12
agencies 178
Assessment Exercise 46–47
centers 106, 122–124, 162–172
organizations 167
support 160, 172
virtual 147
Competency 31, 57, 77, 109–110, 112, 201–205, 224
high level of competence 53, 198, 227
Comprehension 171
skills 144
vocabulary 83
Computer games 146–150, 243
Carmen Sandiego 150
The Sims 2, 150
Conversation 23–24, 41–46, 79, 84, 87, 98, 104, 128, 138–141, 190–200, 254
Countries of origin 163
Cree language 163
Critical Period Hypothesis 20, 57 *see also sensitive period*
Croatian language 231, 237
Cross-cultural understanding viii, 3, 8–10, 156, 159
Cruz-Ferreira, Madalena 222
Cultural
and community centers 166
associations and centers/clubs 167–172
awareness 162–164
connections 166
diversity 168–169, 214
events and classes 52, 165–167, 251
experiences 166
festival 167
values and manners 165

Cybertropism 145
Czech language 122

Deaf 50, 138 *see also American Sign Language*
  community 50
  parents 50, 138
Developmental differences 26, 59, 98, 208, 216, 277
Developmental milestones 204 *see also language milestones*
DeWaele, Jean-Marc 222
Dialects 220–234
Diaz, Junot 5
Disorders 220–224
  developmental 213–214
  learning 214
Diversity 168–169, 214, 271
  Divorce 248–250
  Dolson, David 114
  Dual language 30
  immersion 30, 259–260 *see also two-way immersion*
Dutch language 65, 143–144, 224
DVDs ix, 133, 139–140, 142, 144–145, 149, 151, 261, 135–137 *see also CDs and videos*
  Baby Einstein Language nursery video 28, 135
  Brainy Baby 135
  KidZone 140
  Muzzy series 28, 13
  Plaza Sésamo 139
  TiVo 140

Edutainment 18, 28–33, 133–151, 253
E-mail 123–126, 147, 238
Eliot, Lise 139
Embassies or consulates 172
English as a Second Language 30, 155–156
English language 30, 39, 49, 79, 112–113, 115, 130, 143–144, 154–156, 167, 185, 189, 217, 223, 251
  as a high-status, high-prestige language 19

English language learners 30, 154–158 *see also English as a second language*
English-only policy 217
Error correction 18, 24–26, 33, 128
Errors 22–26, 70–71, 127–128, 174, 213
Expert advice 207–219
Extracurricular activities 69, 170

Families 37–48
  bilingual 78, 196–208
  majority language 100–105
  mixed language 105–110
  monolingual 208
  with more than one child 90 *see also siblings*
Family language audit *see audit*
Family language planning *see family language policies, families*
Family language policies 209, 224
  for multilingualism 224–226
Family language rules *see family language policies*
Family language use (assessing) 199
Farsi language 19 *see also Persian language*
Federer, Roger 243
Feedback 138
Fernandez, Sylvia 208
Finnish language 114
First language 58–60, 83, 99, 116–119, 216 *see also native language*
First language learning 38–40, 83, 115–119, 154–159, 216, 263
First words 11, 26–27, 62, 222
First-born children 78–82 *see also birth order*
Flege, James E. 71
Fluency 20, 22, 91, 112–113, 169, 171, 193, 197–198
Foreign language *see second language*
French language vii, x, 14, 3–50, 143–148, 167, 171, 222, 235–240, 250

Ga language 13
Games 29, 63–68, 74, 78, 98, 102–107, 120, 126, 135, 141–150, 166, 170, 213, 233, 243, 249, 253, 260, 267, 269 *see also activities, computer games*

Gardner, Adrian 103
Gardner, Howard 92–93
Gender 82–89, 255
German language 59, 78, 127, 232
Gopnik, Alison 10–11
Grammar 5–10, 18, 22–24, 37, 50, 54, 60, 65–66, 69–71, 90, 94, 128, 134, 149, 171, 173–175, 193, 196, 213, 215, 231
Grammatical rules 28, 60, 68–70, 75 193–194
and dialects 232
Grandparents 12, 17, 42, 97, 100, 106, 113, 124–126, 146, 190, 217–218, 237–238, 250, 254
Greek language 53

Hall, Geoff 9
Hearing problems 216
Heritage language 11–19, 112, 203, 262
and cultural maintenance 11–12, 13–15
and identity 11–12
and camps 162–167, 182
and classes 167
language programs 162–167, 171
language schools 162–167, 178, 182, 257
High, Pamela 103
Hindi language 39, 49
Hoefnagel-Höhle, Marian 65
Hungarian language 66, 248
Hyperactivity *see Attention Deficit Hyperactivity Disorder*

Identity 3, 72, 267
and bilinguals 12–13
child's 239–241
Imitation of new words 91
Immersion bilingual education 9, 14, 30–31
camps 263
partial 158
programs 153–162, 172
school 164
total 159
two-way 9, 153–162
Immigration 19

Individual differences 56, 72–93
Innate differences 24
Input *see language input*
Instruction 30–31, 69–73, 104, 111, 154–164, 169, 172, 179, 186, 221,
Intelligences (multiple) 92–94
Inter-American Magnet School 160
Internet 106, 126, 144–145, 147–148, 161, 168, 249–250 *see also websites*
iPod 20, 148–149
Italian language 39, 221–223, 230–233, 244

Japanese language 38, 44, 49, 100–101, 240
Job market 3, 13, 52
Job opportunities viii *see also benefits of bilingualism*
Johnston, James C. 51

Kidzone 140 *see also DVDs*
Korean language vii, viii, 10–12, 42–48, 71, 79, 111, 141, 163, 210–211
Verb acquisition 10–12
Krupa-Kwiatkowski, Magdalena 212–213
Kuhl, Patricia 136

LaGasse, Linda 103
Lambert, Wallace 9
Language
density and complexity 56
development *see first or second language learning*
dominant 111, 205, 236
first *see first language*
majority 100–116, 131, 156, 159, 216
minority 81, 109–116, 131, 158, 184, 206
resources 44–45
structure of *see grammar*
support 44, 152
Language awareness (raising) 69, 199
Language choice 45, 54, 89, 115
Language confusion 18, 27, 31, 62, 113–115, 186–190
Language delay 17, 26–33, 62, 113, 184, 207–219
Language drills 174

Language families
  Romance, Germanic 223
Language input 23, 28, 58, 63, 71–75,
    89, 91, 98, 115, 137, 169–170
Language learning *see first language
    learning or second language learning*
Language materials 18, 44, 50, 91, 103,
    107, 135, 148–149, 169–174, 232,
    248–250
Language milestones 58–62, 204–209,
    255, 268
  babbling 59
  one-word stage 60
  two-word stage 60
  vocabulary spurt 60
Language mixing *see code-mixing*
Language Nursery video 28, 135 *see
    also videos*
Language status 19, 243
Language variety *see dialect*
Lanza, Elizabeth 187–188, 196–198
Latin Americans 238
Laws
  Lau v. Nichols 155
  legal restriction 154
  Propositions 227 & 203 154
Lewedeg, Vanessa 107
LeapFrog series 142 *see also activities,
    games*
Learning disability 240
Learning styles 91–92, 169, 269
Learning to talk 136 *see also first
    language learning*
Lesson plans 169
Linguistic Funland 148 *see also web sites*
Linguistic milestones *see language
    milestones*
Linguistic resources 171, 194
Linguistics ix, 17, 161, 210, 215, 252
Listening 129, 148, 150, 171, 174, 177
Literacy 3–7, 14, 67–69, 152, 155–156,
    159–160, 164, 168, 172, 204, 253 *see
    also biliteracy/biliterate*
Liu, Serena H. 71
Lopez, Jennifer 243

Ma, Yo-yo 243
Mainstream classes 157

Majority culture 156, 169
Mandarin Chinese *see Chinese language*
Marchegiano dialect 230
Massey, Douglas S. 112
McCloskey, Jim 231
Memory
  phonological working memory 90
  skills 24
Metalinguistic awareness 5–7, 229
  and academic skills 5
  and literacy 5, 7
  and reading readiness 5
Milestones *see developmental milestones,
    language milestones*
Ming, Yao 243
Minority language 81, 109–111, 113,
    115–116, 124, 131, 158, 184, 206 *see
    also language*
Mistakes 21, 24–26, 57, 61–64, 69, 73,
    126–128, 171, 213 *see also errors*
Mixing languages *see code-mixing*
Monolingualism 19, 30, 158, 207, 217, 245
  monolingual bias 215
  monolingual children 19, 222
    as compared to bilinguals 5, 7,
      62, 208–209
    English speakers 12, 19, 115,
      155, 208
Moore, Sandra 126
Motivation 21, 25, 41, 58, 72–73, 76,
Movies 133–134, 146, 148, 253, 267 *see
    also resources*
Moving (what to do when) 247–250
Multicultural/multiculturalism 8, 15,
    43
Multilingual children 178, 221–222,
    226, 259
Multilingual professionals 13
Multilingualism 14, 220–227
Music 48, 63, 74, 93, 101,142, 144–145,
    148–150, 166, 243, 252 *see also
    activities*
Mutual intelligibility 231

Nannies 52–53, 107, 116, 118–119,
    129–130, 176–177, 222
  Monolingual 177 *see also au pairs,
    babysitters*

Native language 22–23, 30, 39, 61, 115–116, 119–120, 154–157, 159, 164, 237 see also heritage language, first language
Native language models 22–23
Native speaker(s) 18, 22–23, 29, 31, 33, 39, 42, 64, 66, 70–73, 78, 102, 110, 120, 136, 148, 150, 156, 159, 164–165, 170, 173, 226, 230, 232, 246, 248, 250
Native-like accents x, 65, 71–72
Navajo language 163
Nepali language 44
Network 43, 117, 123
New York City 48, 115, 190–191, 193, 250 and Chinese speakers 53
News sources (online) 147
No Child Left Behind Act 30
Non-native speakers 148, 150, 246,
Nonprofit organizations 163–164
Nonstandard language variety see dialects
Non-targetlike language (children's use of) 197–198
Norwegian 187–188, 196–198
Norwegian-Americans 187, 196

Oller, D. Kimbrough 107, 208
One-parent-one-language 18, 27–28, 108, 196, 201, 225
One-word stage 59, 208 see also language milestones

Parenting x, 7, 16, 18, 27, 46, 49, 94, 99, 123, 218, 230, 235–236, 254,
Parenting style 89, 255
Patterson, Janet 104, 138
Pearson, Barbara Z. 111
Pediatricians 17, 216–217
Peers 21, 155, 162, 209, 211, 214–215, 240, 246
  English-speaking 155, 215,
Pen pals 74, 148, 202
Persian (Farsi) 19, 80, 163, 224
Personality v, 24, 77, 84–85, 242, 244, 253, 255 see also temperament
Philosophy
  of teaching/teachers' philosophy/ies 173, 175

of the school or language program 168
Plateau 210, 213
Playgroups ix, 52, 71, 88, 97, 100, 116, 121–124, 225, 248, 250, 254
  language-based 121–122, 124, 213
Playmates 84–85, 122, 190, 225 see also peers
Plaza Sésamo 136, 139 see also DVDs
Polish language 19, 110–111, 120–121, 212
Poplack, Shana 193–194
Portuguese language 19, 49, 53, 143, 156, 222–223, 244
Preferences 9, 14, 43, 91, 93, 109, 122, 197–198
Preschoolers 63, 65, 67
Private language schools 162, 167
Proficiency 204
Pronunciation 65, 67, 91, 188
Psychology ix, 17, 244, 246, 252
Puberty 20–21
Public schools (programs) 72, 154–155, 161–166
Puerto Rico (Puerto Ricans) 190–191, 193, 202

Quadralingualism 223
Questions, open-ended 84

Radio stations (online) 148
  Linguistic Funland 148
  Live 365 148
  online stores, iTunes 148
  Podcasts 133, 148, 150
  Second Life (http://secondlife.com/businesseducation/education.php) 147
  the Voice of America 148
Rate of learning 65
Reading viii, 3, 5, 29, 31, 40–41, 43, 46, 67, 69, 81, 103–105, 114, 121, 125, 128, 133, 136, 138–139, 142, 148, 150–151, 171, 174, 194, 216, 254
  vocabulary 114
Rejection (of a language) 240
Religious organizations 162, 178
Repetition 101, 103, 197

Resources   vi, 20, 41, 43–45, 48, 53, 67,
    70, 76, 87, 116, 134, 145, 147, 151,
    153, 155, 161, 163, 165–166, 168,
    171–172, 178–179, 216, 218,
    247–250, 253 *see also activities*
Rituals   103
Rochester Chinese School   167
Role models   247, 255
Rolstad, Kellie 158
Rosenbusch, Marcia   179
Roy, Arundhati   5
Rumbaut, Rubén G.   **112**
Rushdie, Salman   5
Russian language   viii, 45, 47, 49, 53,
    142, 146, 163, 194, 203, 250–251

School districts   153, 157, 161, 210
School language   8, 43
School-aged children   29, 68–69, 71–74,
    79, 143, 145, 148–150, 153, 164, 246
  and analytical skills   69
  and correction in foreign language
    learning   70
  and learning styles   70
  social and emotional development   73
  and second language learning   29
Second language acquisition *see second
    language learning*
Second language learning   vi, ix, x,
    17–18, 20–22, 26, 33, 44, 52, 58, 73,
    77, 82, 84–85, 87–90, 94, 97,
    99–100, 110, 121, 130–131, 133, 135,
    146, 148–150, 152–153, 238, 248,
    252–255
  activities/materials   92, 139, 152 *see
    also activities*
  and accents   x, 22, 55–58, 65–73
  and adults   21–23, 59
  and age   21–22 *see also age*
  and aptitude   89–90 *see also aptitude*
  and attitudes   71
  and correction   72
  and emotional pressure   59
  and fluency   24
  and grammar   24
  and identity   74
  and incentives for learning   48
  and individual differences   58

  and learning styles   72
  and mistakes   63
  and preschoolers   65–70
  and programs   vi, 8, 14, 152 *see also
    bilingual education programs*
  and pronunciation   65–67, 91, 188
  and rate of learning   67
  and reading   69
  and self-esteem   75
  and mistakes   27
  and older children   23
  and reading   69
  and siblings   25–26
  and ultimate level of success   23
  and ultimate success   23
  as a lifelong process   207–210, 253,
    259–260
  at home   97
  classes   46–50, 71, 91–94, 104, 124,
    130–131, 163–178
  contextual clues   171
  incentives for   45–48, 54
  knack for   89
  optimal conditions/strategies for   17,
    70–73, 84–89, 97, 101, 241, 250
  technology, assisted   133, 139–150,
    243
Self-confidence   15, 128
  consciousness   44, 69, 246
  esteem   viii, 3, 12, 15, 73, 166,
Senesac, Barbara   160
Sensitive period   57–58 *see also Critical
    Period Hypothesis*
Separation (of languages)   28, 33,
    200–201
Serbian language   231
Sharapova, Maria   243
Shin, Sarah   79
Siblings   23–24, 63, 78–82, 137
Silent period   211–213, 219
Singapore   13, 31, 43, 221–223
Skehan, Peter   89
Skutnabb-kangas, Tove   114
Snow, Catherine   65, 115
Social prestige (of dialects)   229
Society's norm (multilingualism as)   225
Software   vii, 133–134, 149–150 *see also
    computer games*

Sounds (learning) 5, 38–39, 59, 65–67, 91, 137–138, 142, 185–186, 189, 211
Sound systems 66, 91, 189
Spanish language vii, viii, x, 7–10, 14, 19, 24, 31, 37–38, 41–45, 47–49, 52–53, 86, 100–102, 104, 107, 111–114, 119, 121–122, 129–131, 133–134, 136, 138–141, 143, 152, 154–156, 158–160, 162, 184–185, 188–189, 191–193, 200, 202–203, 205, 208, 214, 223, 226–228, 238, 241, 243–247
    as unofficial second language of the United States 52
    and which language 37–38
Spanish-English bilinguals 52, 193
    and the job market 52
Spanish-English dual immersion 31, 160, 162, 238
Special needs, children with 209, 216–218
Specific language impairment (SLI) 216
Speech therapists *see speech-language pathologists*
Speech-language pathologists 82, 209, 217
Speed of learning 21
Stabilization 213
Standardized tests 115, 160
Strategies 87, 160, 189, 197–198, 212, 218, 228, 245, 251, 255
    children's 189, 197, 228
    one-language 198
    parents' 197–198
    social and cognitive 87
Subtitles 30, 134, 144, 150
Summer camps 30, 44, 69, 162
Swain, Merrill 171
Swedish language 109, 114, 144–145, 210, 222–223
Swiss-German language 232
Switching languages *see code-switching*
Syllables 91

Tagalog language 19, 47, 203, 226–227
Taiwanese language 53, 111, 164, 236–237
Talking toys 18, 28, 134–135, 140–141

Tamil 221
Target language 163–164, 172, 198, 213, 253
Tawada, Yoko 5
Teachers vi, 4–5, 17–18, 22, 25, 27, 29, 43, 61–62, 70, 72, 111, 113, 118–119, 138, 152–153, 155, 159–160, 163–164, 169–180, 194, 202, 211, 214–216, 240, 242, 253
    experience 174
    as native speakers 170, 173
    observation 175–176
    and portfolios 175–176
    private 178
    qualifications of 170
    retention of 170
Teaching approach/philosophy 173, 175, 177
Teenagers 14, 38, 43, 117, 228
Television/TV 7, 16, 29, 129, 134–144, 146–148, 151–154 see also DVDs
    video games 147
    Big Bird 142
    Dora the Explorer 20, 133, 139, 143
    Elmo 139, 141–142
    Go Diego Go 139
    movies 133–134, 146, 148, 253
    Muzzy 28, 135,
    Plaza Sésamo 136, 139
Tests viii, 4–5, 65, 93, 115, 160, 231
Third languages 41, 99, 144, 225, 228, 255
Tivo 140 *see also DVDs, Kidzone*
Toukomaa, Pettri 114
Toys, Bilingual vii, 18, 20, 28, 57, 63–64, 68, 82, 118, 123, 134–135, 137, 140–145, 149–150 *see also talking toys*
Transitional bilingual education *see bilingual education*
Trilingual 5, 28, 135, 217, 221, 223, 224–225, 230, 233, 252
Trilingual language learning 31
Trilingualism vi, 32, 220, 224, 233–234
Tuition 163
Tutors 44, 130, 172, 176, 178, 253 *see also teachers*

Twin Cities Chinese Language School 163

Two-way immersion 9, 30, 153, 159–162 *see also dual language immersion*

Two-word stage 59–60, 208, 222 *see also language milestones*

U.S. Census Bureau 47, 245

U.S. Department of Labor Bureau of Labor Statistics 247

Ukrainian language 61–63

Ultimate level of success 21–22

United Nations International School in New York City 115

United States 149, 154–156, 158, 161, 163, 167, 198, 202–203, 207, 209, 211–212, 215–217, 220, 221, 245, 251

Urdu language 222

Valencia, Jose 228

Variation (in language development) 26, 56, 59, 62, 208–209

Video games 143, 145–147, 150, 243 *see also resources*

Japanese baseball video games 147

Video(s) 28–29, 126, 129, 135–137, 145, 148–150, 174, 222, 233, 253 *see also resources, TV*

Baby Einstein 135

Thomas the Tank Engine 129

Vietnamese language 163

Vocabulary 18, 24, 27, 54, 60, 65–67, 70, 78, 82–83, 86, 98, 103–107, 114, 118, 128, 134, 138–139, 141–142, 145, 149, 187, 208–209, 211, 214, 231, 253

basic 103

checklist 83, 139

growth of 98

items 145

learning 149, 211

production and comprehension of 82–83

productive 107

size of 104, 138–139

skills 104

words 27, 60, 83, 104, 196

Vocabulary spurt 60, 187, 208

Washington International School 14

Washington Japanese Language School 165

Web sites 137

Amazon.com (http://www.amazon.com) 259, 260

American Academy of Pediatrics, Television and the family (http://www.aap.org/family/tv1.htm) 136, 275

American Fact Finder (http://factfinder.census.gov/) 46

The Bilingual Edge (http://www.thebilingualedge.com) 167, 216, 259

Bureau of Labor Statistics (http://www.bls.gov/) 277

Directory of Chinese language programs in the U.S. (http://www.internationaled.org/uschina/K12ChinesePrograms.xls) 161

Directory of two-way bilingual immersion programs in the U.S. (http://www.cal.org/twi/directory/) 161, 276

Languages of the World (http://www.ethnologue.com) 278

Fairfax County Public Schools (http://www.fcps.edu/DIS/OHSICS/forlang/partial.htm) 161

Foreign language immersion programs in the U.S. (http://www.cal.org/resources/immersion/) 161

Grandcare: Encouraging that special relationship between your parents and your kids (http://www.washingtonparent.com/articles/0411/pep.html) 275

Guidelines for starting an elementary school foreign language program (http://www.cal.org/resources/digest/rosenb01.html) 179, 276

Children's personality portraits (http://www.personalitypage.com/kid_portraits.html) 268

Linguistic Funland (http://www.tesol.net/penpals/) 148

ABC's of child development (http://www.pbs.org/wholechild/abc/)

National Clearinghouse for English Language Acquisition (NCELA) and Language Instruction Educational Programs glossary (http://www.ncela.gwu.edu/expert/glossary.html) 156, 276

Net Nanny™ (http://netnanny.com) 146

Second Life (http://secondlife.com/businesseducation/education.php) 147

Shutterfly (www.shutterfly.com) Two-way bilingual immersion: A portrait of quality schooling (http://brj.asu.edu/content/vol26_nol/html/art6.htm) 276

Individual Profiles for child/ren (http://www.uniquelyyou.net/child/index.php) 268

U.S. Census Bureau, American community survey (http://www.census.gov/acs/www/) 272, 277

You Tube Broadcast Yourself™ (http://www.youtube.com) 146

Welsh language 44

West Ulster English 231–232

Which language v, ix, 19, 27–28, 32, 35, 37–38, 40, 43–48, 52–54, 61, 128, 162, 186, 188–190, 196, 206, 255,

Word boundaries (learning) 211

Word choice 239

Writing ix, 5, 41, 53, 123, 171, 174, 205

Xhosa language 39

Yeni-Komshian, Grace H. 71

Zentella, Ana Celia 190–191